Global Congress on Process Safety 2021

AA008999

Topical Conference at the 2021 AIChE Spring Meeting and 17th Global Congress on Process Safety

Online
18-23 April 2021

ISBN: 978-1-7138-5731-0

Printed from e-media with permission by:

Curran Associates, Inc.
57 Morehouse Lane
Red Hook, NY 12571

Some format issues inherent in the e-media version may also appear in this print version.

Printed with permission by Curran Associates, Inc. (2022)

For permission requests, please contact AIChE
at the address below.

AIChE
120 Wall Street, FL 23
New York, NY 10005-4020

Phone: (800) 242-4363
Fax: (203) 775-5177

www.aiche.org

Additional copies of this publication are available from:

Curran Associates, Inc.
57 Morehouse Lane
Red Hook, NY 12571 USA
Phone: 845-758-0400
Fax: 845-758-2633
Email: curran@proceedings.com
Web: www.proceedings.com

TABLE OF CONTENTS

621490 The latent triggers of disasters are alive - deep inside your company

Abstract:

Never in our history have our workplaces been so safe and our workers so well protected. Yet in recent years, something strange started to happen. While many companies have 'hit a wall' in safety performance, some have seen a slow but visible increase in serious accidents, and most alarmingly, a few 'best-in-class' companies suffered catastrophic events. It seems as if the closer we get to 'zero accidents', the more vulnerable we become for serious events to happen... Something is wrong. Making the systems safer, puts the organization at higher risk? This presentation will demonstrate how the few seconds before a disaster were years in the making – by well-intended, but dangerous ideas, beliefs and delusions that become destructive forces in the culture. It will outline the new roles of leaders and practitioners in a different world awaiting us.

HIRA in Mining Companies Is Necessary and a Big Challenge

Presenter/Principal Author
Américo Diniz Carvalho Neto
RSE CONSULTING
Brazil
americodcn@rsem.com.br

Other Author(s)
Afonso Costa
RSE CONSULTING
Brazil
Afonso.costa@rsem.com.br

Prepared for Presentation at
American Institute of Chemical Engineers
2021 Spring Meeting and 17th Global Congress on Process Safety
Virtual
April 18 - 22, 2021

HIRA in Mining Companies Is Necessary and a Big Challenge

Presenter/Principal Author
Américo Diniz Carvalho Neto
RSE CONSULTING
Brazil
americodcn@rsem.com.br

Other Author(s)
Afonso Costa
RSE CONSULTING
Brazil
Afonso.costa@rsem.com.br

Keywords: HIRA Project,

Abstract

The implementation of a risk-based process safety management system is a turning point for any company that intends to have a better understanding of the risks associated with their process. However, full implementation of all elements takes time, and in the meantime, catastrophic scenarios should be managed accordingly, without wasting time.

To fulfill both criteria, having long and short run implementation strategies, HIRA (Hazard Identification and Risk Analysis) process was developed for mining companies worldwide. Based on the Critical Control Management (CCM), HIRA process is focused on "quick wins" of process safety, management gaps that must be eliminated shortly, to keep high consequences scenarios from happening. Those scenarios are known as MUE (Material Unwanted Event).

HIRA for mining companies is based on the methodology developed by ICMM (International Council of Mining and Metals), which is comprised of 9 steps, 6 of them are only to identify scenarios and their causes, consequences and mainly the critical controls associated. The methodology intends not only to identify critical controls, but also to check their status (implemented or not implemented), identify items that can make them more reliable and to assign responsibility for the management of the control, throughout its lifetime, to guarantee they will perform their function when demanded.

To accomplish such tasks, HIRA process widely uses Bow Tie Analysis, a very intuitive and visual risk management tool, which allows even people who do not have a lot of experience in the subject to easily understand the graphic representation of the MUE, such as cause, consequence, top event and mainly the controls and their status. Bow tie sessions are followed by field verification of each critical control identified, to check coherence between records (data), integrity and reliability, of the controls.

As a result, both existing and possible controls (known as Future State) are identified, according to the MUE analyzed and, comparing information from field verification, a qualitative risk assessment is carried out for three different categories – Inherent, residual and forecast risk.

By the end of the HIRA process, many types of controls are identified, such as procedures, safety valves, inspection plans, guard rails, alarms, interlocks…and so on. They now will be managed differently, as they are classified as critical controls – items that are crucial do prevent the MUE form happening or mitigate its consequence.

1 Main section 1 heading (Introduction)

Many companies have improved their performance on safety-related issues using a wide range of rules, procedures, and standards; however, Material Unwanted Events, such as fatal accidents, rare catastrophic events and circumstances of over-exposure that are harmful to health, are still subject to occur. For this reason, the ICMM (International Council for Mining and Metals) developed an approach called Critical Controls Management (CCM), aiming at the continuous improvement of management over rare events, but potentially catastrophic, emphasizing critical controls. The methodology to improve the management control of these types of events, called Material Unwanted Events (MUEs) follow below:

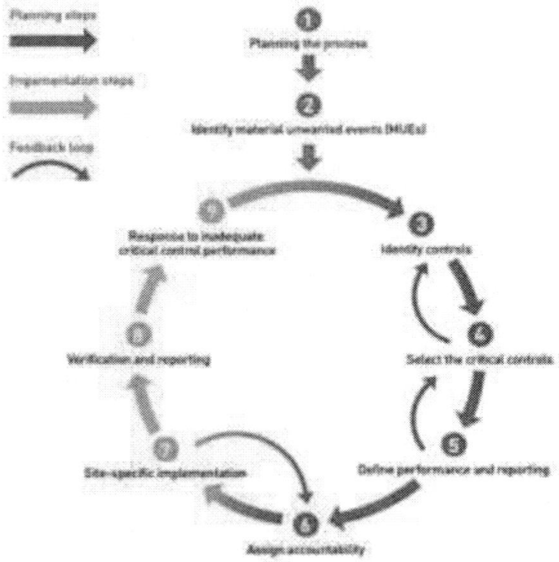

ICMM Critical Controls Management Process (CCM)

CCM's approach is based on some STEPS as follow:

Step 1: Develop a project plan that contemplates the organizational context, objectives, team responsibility and the individuals and business areas affected in the implementation.

Key actions for STEP 1

• Develop a plan that describes the scope of the project.

This includes:
– organizational context
– project objectives

– responsibilities
– business sections involved.

• Develop methods to:

– identify potential hazards and unwanted events
– assess risk
– review MUEs
– select critical controls
– assess objectives and performance of critical controls
– investigate critical control underperformance
– measure impact of the project
– identify ownership and accountability.

Step 2: Identify the MUEs to be managed and summarize the main information for each MUE. In order to identify, verify if they present material risk, evaluate the opportunities to eliminate it by improving the designs and describing the MUE completely (hazard, materialization mechanism and possible consequences).

Key actions for STEP 2

• Understand major hazards and identify potential MUEs.
• Apply selection criteria to MUEs with a focus on the consequences.
• Identify design opportunities to address the hazard, reducing the potential consequences and eliminating the MUE from the CCM process.
• Describe the identified MUE, including the relevant hazard, mechanism of release and nature of the consequences.

Step 3: Identify the controls, developing Bow Ties diagrams and evaluating them.

Key actions:

• Identify the controls.
• Prepare a bowtie diagram.
• Assess the adequacy of the bowtie and the controls.

Step 4: Select the critical controls for each identified MUE, summarizing its main information.

Key actions For STEP 4

• When identifying critical controls, apply the critical control definition and guidance in this section.
• Consider the performance requirements of the potential critical controls and how they could be verified.

• The final set of critical controls for an MUE should represent the critical few that, when managed using CCM, can effectively manage the MUE risk.

Step 5: Define the objectives and Critical Controls performance requirements, how they will be checked and the activities that will be affected by them.

Key actions for STEP 5

• Define objectives and performance requirements for each critical control.
• Identify current activities that affect the critical control's performance.
• Describe activities to verify performance and reporting requirements.
• Identify what would trigger immediate action to stop or change the operation and/or impose the performance of the critical control.

Step 6: Develop a list of those responsible for the MUE, for the Critical Control and for its verification, including the preparation of the final information report.

Key actions for STEP 6

• Assign owners for MUEs, critical controls and verification activities.
• Describe reporting plan for the health of critical controls.
• Assign owner for review of reports.

Step 7: Define a verification plan and reporting plan for each MUE, implementing it according to the specific needs.

Key actions for STEP 7

• Critical control information must be specific to a site or asset.
• Adjust the critical control definition, performance information and verification requirements as necessary to suit the local context.
• Site-specific planning for implementation may involve an iterative process.
• Site-specific planning should include establishing a foundation for CCM that includes leadership, communication and appropriate development of knowledge and understanding related to the critical controls.

Step 8: Implement verification activities and report the process, defining the status of each critical control.

Key actions for STEP 8

• Undertake verification activities for critical controls as described in MUE/critical control verification and reporting plan (developed in Step 5).
• Report a summary of verification activity results to the critical control owner.

• Report critical control verification status to the MUE owner.
• Reports should highlight priority information succinctly using traffic light system.
• Action initiated if critical control performance drops below the defined triggers (established in Step 5).

Step 9: Performance investigation if critical controls are performing below expectations.

Key actions for STEP 9

• Take action when critical control performance is inadequate (below the defined trigger threshold).
• Investigate the causes of unacceptable critical control performance.
• Information and data from the investigation should be used to continuously improve the CCM.

The companies risk map, in most cases, does not fully reflect the main risks to which the operation and the external population are exposed. This is due to the lack of a unified methodology to identify, quantify and measure risks. In view of this reality, one company can initiate the HIRA Project (Hazard Identification and Risk Assessment), to conduct a quick technical scan of risks in its operations to identify relevant and unmitigated situations.
HIRA was designed to quickly identify and verify the main risks of the company units operations. This is a global project to be implemented and will be at the center of the operational risk agenda in the coming months. The HIRA Project will focus on the very critical risks of severity in the company operations and its subsidiaries and its results, will be reviewed frequently by the Executive Committee for Operational Risks, the Executive Board and the Board of Directors. His results will guide the main actions to reduce the organization's risk profile.

GOALS

The current study aims to identify and analyze operational safety risks with a very high-risk magnitude or very critical severity of a company site. In addition, the identification, definition of performance criteria and the establishment of criteria to guarantee the performance of associated critical controls are part of the scope of this work. From this, it will be possible to help the company to produce a complete and accurate inventory of the relevant unwanted events of the organization (MUEs) and, later, based on the identified MUEs, identify critical controls, associated performance requirements and means of verification.

For this purpose, this study will include:

1. A consolidated list of operational safety risks assessed with a very high risk magnitude or very critical severity (MUEs) for the unit;

2. Bow Ties based on the identified MUEs and a corresponding list of critical controls identified for each (following the suggested ICMM decision tree and based on the group discussion);

3. One-Pager for each critical risk control.

4. List of action plans to be developed based on the prioritization of risks.

5. Updated MUEs risk matrix of the unit.

METHODOLOGY

The Bow Tie Analysis (BTA) can be used to identify and assess hazards with the potential to cause major accidents involving company site. This study can be conducted through presential or remote meetings through the TEAMS videoconferencing platform, by multidisciplinary teams composed of corporate company members and company site members and can be supported by external consultants.
Initially, happens a Workshop session, jointly between the consultants of RSE Consulting and company employees present at the meeting. The site MUEs were defined taking into account that the chosen MUEs would be among the most at risk in the facilities analyzed. In these sessions, MUEs were selected and evaluated from a previous list of more than 170 different MUEs, from different units of the company, including units in different countries which the HIRA project will be developed. Other new scenarios were also discussed and included as discussed with the local technical team. After determining the likely MUEs applicable to the site, these were then categorized as follow:

• Category 1: Scenarios in which, all the information presented during the Workshop, was sufficient for validation as a MUE. These are considered priority scenarios;

• Category 2: Scenarios in which, despite the information presented during the Workshop, these were not sufficient to validate the MUE in remote mode, thus requiring field visits and more information for validation. Also, in this category are included scenarios that can be considered as developments of scenarios classified as Category 1;

• Category 3: Scenarios in which, despite the information presented during the Workshop, these were not enough for validation as a scenario that meets the HIRA criteria, to be considered MUE. An on-site visit would be necessary to complement information and analysis. Thus, the team present at the MUE Workshop session with different specialists concluded in a list some (%)

MUEs classified as Category 1, that is, priority to the units of the complex. These meetings can be remotely, and the list was
subsequently validated by company leaders. It should be noted that the identification of MUEs needs to consider the historical events (Accidents and Near Miss) of the site, company and sector, in addition to predictable events, given the operations and activities at individual locations. Therefore, the identification of MUEs needs to include personnel with adequate experience and a review of relevant data.
Then, meetings were held remotely to analyze the MUE and generate project deliverables for the Complex. The analyzes were composed by members of the company corporate members, site regional members and external RSE consultants, where the consequences, threats and the respective existing controls for each MUE of the site were identified. During these meetings Bow Ties from other company units were analyzed with similar processes as a reference. Details about the Bow Tie technique are presented of this paper.

To identify the critical controls and controls of the MUEs, the proposal decision tree
was followed:

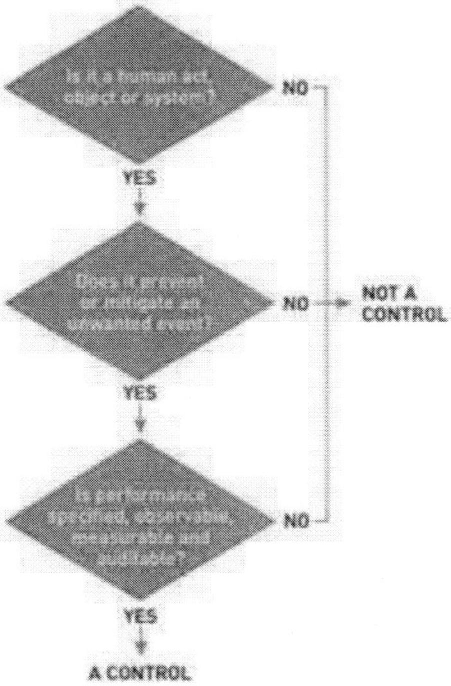

Parameters for defining a control

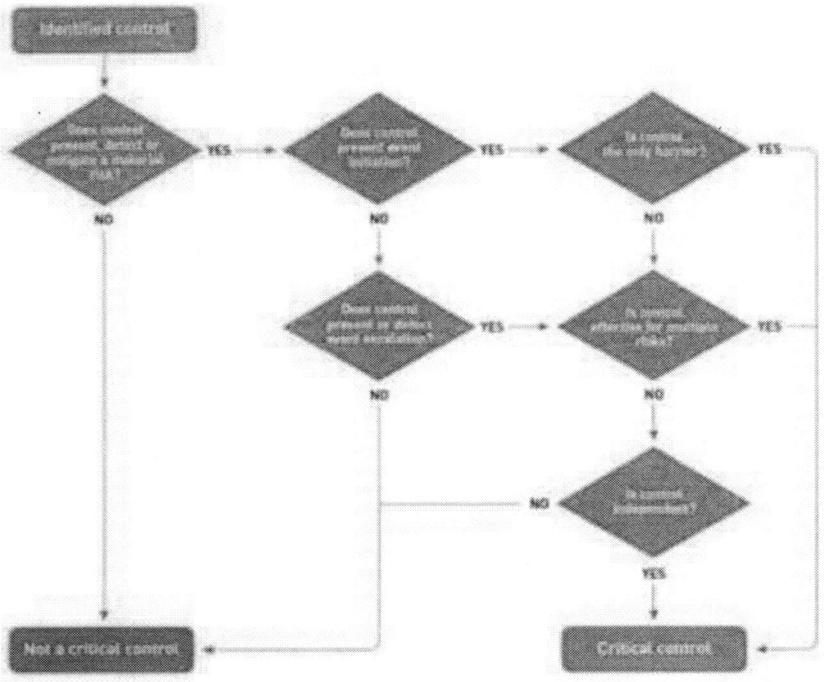

Parameters for defining a Critical Control

Once the critical controls were identified, a "One Pager" was generated for each one, that is, a quick briefing of just one page for each critical control (except "Future State" critical controls) containing the following information:

- Objectives of related critical control
- Performance requirements to meet objectives
- Activities that can enable or strengthen control
- Activities that can be verified to attest to the performance of a critical control
- Performance for critical control
- Trigger to stop the unit, revision or investigate, depending on the performance of the control
- Critical Controls attributes

The current levels of effectiveness (status) of performance of these critical controls were classified initially without field verification, based on the opinion of company specialists present at the meetings, to generate previous actions for the company, according to the methodology established for carrying out remote HIRA, during the COVID-19 pandemic. However, these classifications must be rectified / ratified in another phase of HIRA, when the field visit is done.
Remote risk classification meetings were also held, categorized as Inherent, Residual and Forecast, classifying them as follows:

- Inherent risk: premise = the existing critical controls to prevent each cause or mitigate each consequence are null (there are no controls);

• Residual Risk: premise = the existing critical controls to prevent each cause or mitigate each consequence are fully implemented or not implemented, depending on the classification of the status of each critical control determined through remote meetings by the Vale specialists team

• Forecast Risk: premise = critical preventive controls, mitigating controls, future controls (Future State) and all actions and recommendations captured during the meetings are fully implemented and functional, that is, they are compl ete and reliable.

It is important to note that risk assessments were carried out, remotely, without the field verification by the HIRA teams to validate, based on the One Pagers, the critical controls. Therefore, critical controls were evaluated in future scheduling, considered the next phase to be carried out at HIRA on site.

For the risk analysis of each MUE, the risk acceptability criteria of the company must be adopted. Below follows a brief explanation of Bowtie methodology selected in this Project.

BOW TIE TECHNIQUE

Bow Tie is a qualitative risk assessment method used to assess how high-risk scenarios develop. The essence of Bow Tie consists of plausible risk scenarios surrounding a certain hazard, and ways in which the organization prevents these scenarios from happening. The method takes its name from the shape of the diagram you create, which looks like a male bow tie.

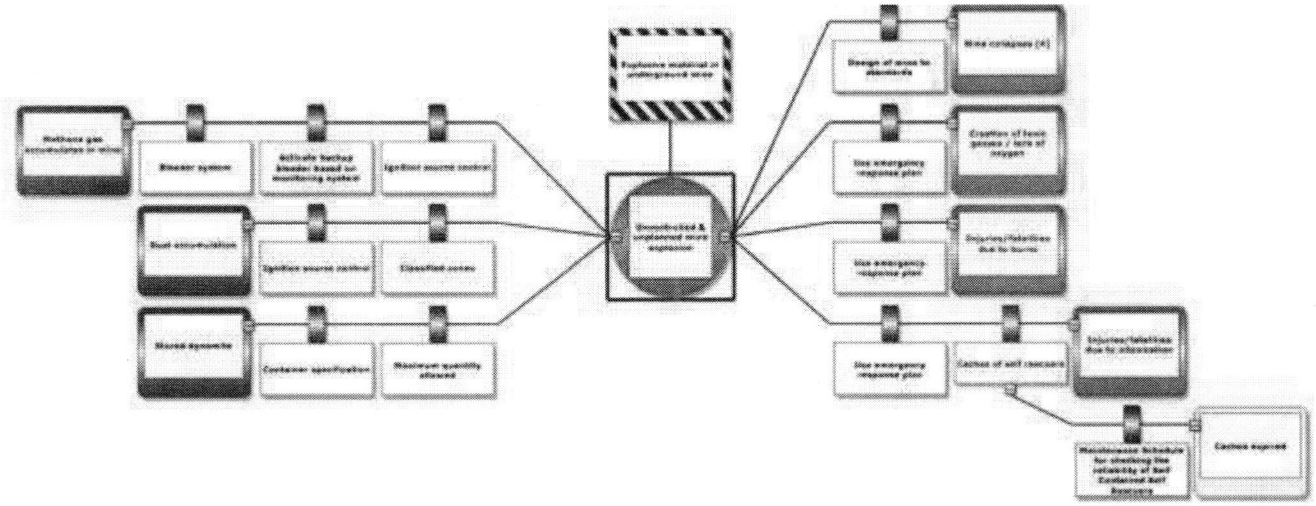

The Bow Tie methodology has several objectives:

• Provide a framework for systematically analyzing a hazard.

• Helps to decide whether the current level of control is sufficient (or, for those who are familiar with the concept, whether the risks are as low as reasonably practical or ALARP).

• Helps to identify where and how to invest in resources that could have the greatest impact.

• Increases communication and risk awareness.

General step by step construction of the Bow Tie diagram:

Hazards Identification The purpose of hazard identification is to find, recognize and describe the risks that can help or prevent an organization from achieving its goals. Hazards can be operations / activities (operating rotating machines, driving a car),

I. substances (chemicals, hot fluids, etc.) or situations (a suspended load) that we deal with in the normal processes of our business. As long as these hazards are under control, they will not cause harm, but introduce the potential for harm.

II. Definition of the Top Event / MUE When control over a hazard is lost, it is usually possible to identify the exact moment when a normal situation changes to an abnormal situation, that moment is called Top Event. The Top Event is not yet a catastrophe, but the company is now exposed to the potential for harm from hazard.

III. Identification of Consequences The consequences are classified as the possible negative results directly linked to the top event, which can affect people, the environment, the facilities and the company's reputation.

IV. Threats Identification Factors that can lead directly to the top event are called threats. These are capable of causing the Top Event independently and are not associated with barrier failures.

V. Identification of Preventive Controls Risk management is about controlling risks. This is done by implementing barriers to prevent certain events from forming. A barrier (also called control) can be any measure taken that acts against some unwanted force or intention, in order to maintain a desired state. Preventive barriers, identified on the left side of the Bow Tie, can prevent the top event from happening or can eliminate the threat.

VI. Identification of Recovery Controls Controls or recovery barriers identified on the right side of the Bow Tie can prevent the top event from resulting in unwanted consequences or mitigate the consequences.

VII. Define Escalation Factors Once the controls have been identified, it is recommended, according to the Bow Tie method, to identify specific conditions or actions that reduce the effectiveness of the control. These are called escalation factors.

VIII. Identification of Escalation Factors Barriers Once the escalation factors have been identified, the last step is to look at what barriers we have to manage these factors. These barriers have the same concept as all the barriers previously discussed, but, now they do not prevent / mitigate a top event or consequence, they will prevent a barrier from failing. The BowTieXP software, version 9.2, developed by CGE Risk, was used to construct the Bow Tie diagrams. BowTieXP is the most widely used risk assessment software based on the Bow Tie methodology that allows you to easily create risk assessment diagrams.

ANALYSIS RESULTS

Table attached shows the Material Unwanted Events (MUEs) that were hipotetical identified and jointly defined by RSE Consulting and the company team. As defined by the company, for an event to be considered as MUE it must have the potential to cause one or more fatalities or the potential to cause a major environmental impact that goes beyond the limits of the company.

In addition to these available scenarios (Category 1), the MUEs Workshop, conducted with a multidisciplinary team, generated potential scenarios for MUEs whose information presented was not sufficient to validate them in the remote mode, requiring additional information through field visits for validation (Category 2 and 3).

In total, 221 existing controls were defined to prevent and / or mitigate MUEs and 138 of these were classified as critical controls for these scenarios. These numbers consider the proposed future controls and the repetition of the same control between the scenarios.

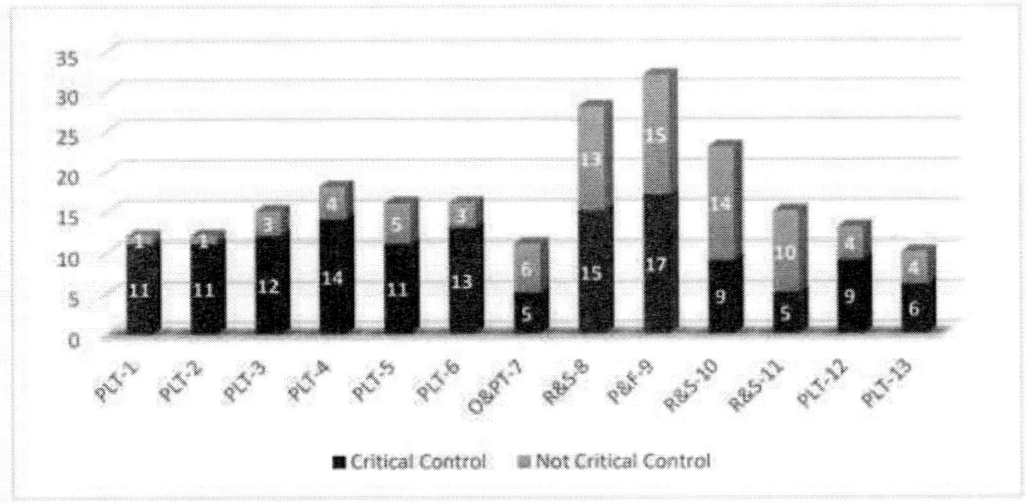

Controls and Critical Controls

According to the criteria adopted for the status of critical controls, the following were identified:

✓ 98 Critical controls classified as "Implemented"

✓ 09 Critical Controls classified as “Not Implemented”

✓ 31 Critical Controls classified as “Future State”

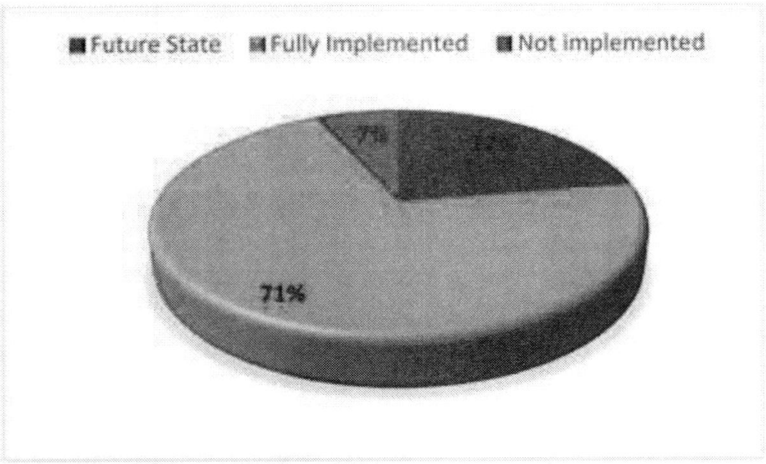

Percentage distribution of the status of Critical Controls

Regarding the type of critical controls, the following were identified:

✓ 104 Critical Controls classified as "Preventive";

✓ 34 Critical Controls classified as "Mitigatory".

This distribution demonstrates the greater effort of the professionals who participated in the assessments and identification of risks / controls in the Search for preventive solutions instead of mitigating unwanted scenarios, which satisfies the initial premise of HIRA.

Additionally, considering a larger sample of 19 sites, where 252 MUEs were analyzed, one can have a better understanding of the results.

• Critical controls - for the 252 MUEs evaluated in this sample, 4283 controls were identified, of which 2487 were considered critical controls. In this number are included the controls considered Future State, that is, the proposals identified by the group to reduce the risk of the assessed scenario. This means that 58% of the controls were considered crucial in preventing the occurrence or mitigating the consequences of the MUE, according to the evaluation team. This is always a point of attention when evaluating controls because it is not uncommon for members to tend to classify a large percentage of controls as critical. This is because companies prioritize resources to ensure the reliability of their critical controls, and the Operation team can use this criterion as a factor for prioritizing resources, going in the opposite direction to the methodology, which is a big mistake. It is up to the consultant to ensure that the flow of classification of critical controls is properly followed.

A rule used, which has served for the present project, is to consider as critical controls something between 45% to 55% of the controls. If the numbers indicate something greater than this, a ratification with the evaluation team is recommended.

For the same project, 1115 controls considered critical were classified as administrative controls, that is, they are basically controls that are highly dependent on human action to fulfill their required function. Examples of these controls are procedures and inspections. The latter is one of the main points of attention because it presents many opportunities for improvement as inspections are common not only in mining but in any company. However, many operational inspections carried out do not have the required reliability, that is, they are not able to identify deviations in integrity that can lead to the MUE.

These inspections are characterized by having a high frequency (sometimes daily) of execution and low level of details. On the other hand, there is the integrity inspection, which has a low frequency and a high level of details. The latter is usually performed by certified inspectors, has an associated procedure and inspection report issued at the end of each inspection. That is, depending on the type of inspection carried out, it is possible that this process is very effective in identifying structural damage or just a simple routine daily verification process, without the required level of detail.

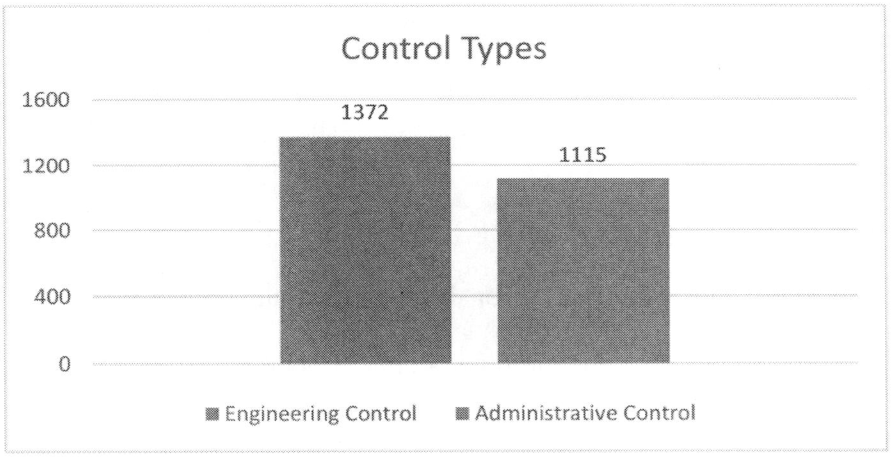

Graph 2 – Types of Critical Controls

It is highly recommended that this information about the items that can enable or strengthen critical control, in the case of operational inspection, are clearly described in the One Pager.

Preventive or mitigating controls - Considering the type of control as preventive or mitigating, again for the project considered, it was found that 1874 (75%) were preventive and 613 (25%) were mitigating. The mitigating controls are basically the emergency action plan and fire fighting system. Less frequently, dykes also appear as important controls for the MUE of loss of fuel containment.

It is important to say that, although they represent only 25% of the identified critical controls, mitigating controls are responsible for a higher percentage of recommendations. One of the most recurrent is the failure to consider the MUE in emergency pre-plans and drills.

Conclusion:

Throughout HIRA´s remote work of the company site, risk scenarios considered more critical were evaluated, with the potential for multiple fatalities or significant environmental impact or level 1 process safety event. Although the result of critical controls considered as implemented was in higher percentage, deviations were identified in the management of some pillars of Process Safety, such as Mechanical Integrity, and Risk Management. For a better understanding of the deviations listed, the list of actions is done.

A larger number of "automatic" or engineering critical controls type were identified in relation to those with human interface, such as operational procedures, inspections, maintenance plans, among others. For some of these controls, actions were proposed to ensure greater reliability on demand, activities such as systematic inspections, maintenance, calibration plan, looping control tests, fail-safe mode, etc.

About the critical controls identified with human actions, it is necessary the following requirements to the reduction of the failures: to reach deadlines of procedure's review, elaborate training plan/program with knowledge retention (evidence with records), recycling and periodic evaluation of the team performance, with actions for "under the established standard", as well as investigation records of near miss incidents.

About company site, it is important to highlight:

1. Treat the critical controls as critical items into the SAP or the specific management software.

2. Ensure proper inspection and maintenance for the automatic controls (safe interlock systems);

If the field verification of critical controls is performed in the company site made it possible to diagnose opportunities for improvement, strengthening and implementation of Process Safety Risk Management. Considering the concepts of hazard identification, risk analysis, change management, implementation of Process Safety Management System will be essential for the company site to manage its risks associated with the MUEs, to keep them at an acceptable level.

2 References:

[1] RSE Consulting HIRA methodology process, Americo Diniz Carvalho Neto, 2017.

[2] RSE Consulting Barriers Audit methodology, Americo Diniz Carvalho Neto, 2007.

[3] Health and safety critical control management Good practice guide, ICMM.

[4] CCPS. *Guidelines for Bowtie in Risk Management.* Center for Chemical Process Safety, American Institute of Chemical Engineers, New York, NY, 2018.

[5] CCPS. *Guidelines for Hazard Evaluation Procedures.* Center for Chemical Process Safety, American Institute of Chemical Engineers, New York, NY, 2008.

616590 Title: Silencing a Noisy Agitator Seal By Managing Stress

Abstract:

Title: Silencing a Noisy Agitator Seal by Managing Stress

Abstract: Did you know the sound that comes when you run your finger on the rim of a wine glass is referred to as "partial boundary layer lubrication"? Did you know this same phenomenon also occurs commonly on buffer and media tank agitators in bio-processing facilities? However, on agitator seals the sound can amplify to over 90 decibels, which exceeds OSHA's personal exposure limits.

Using an FEA model, John Crane found, that the noise from partial boundary layer lubrication is a result of vibration caused by the application of an even or consistent stress to the seal face, which is consistent with field trials since the seals rarely squealed at startup. Instead, the seals squeal after some run time when the right amount of stress is applied. This stress could come from the pressure of the vapor space in the tank or vessel, which also explains why the noise level and pitch can come and go depending on the pressure in the tank, agitator speed, as well as other scenarios.

This led to the discovery that the squealing can be reduced, or even eliminated, by managing the stress imposed on the agitator seal faces. Subsequently, John Crane developed a solution that results in an active seal face, meaning the degrees of flatness change with the stresses applied during operation. In other words, the rim of the wine glass changes as you run your finger around it preventing the noise condition from ever taking place.

Take Home Message from Poster: The noise from a media or buffer tank agitator seal can be reduced, making a quieter and safer workspace.

619172 Use of Data Analytics to Obtain Cost Reductions and Performance Improvements in a Regulatory Application

Abstract:

All operating nuclear power plants (NPPs) in the United States are required to monitor and assess the performance of plant structures, systems and components (SSCs) as specified in the Maintenance Rule (10CFR50.65). This rule requires that all licensees identify the occurrence of any SSC failure that results in a loss of safety function, often referred to as a Maintenance Rule Functional Failure (MRFF). All MRFFs are subsequently evaluated through the plant corrective action program (CAP) as required by 10CFR72.172. To ensure these requirements are met, all NPPs document events that occur which are evaluated to identify impact on the plant. A typical dual unit generating facility generates as many as 15,000 of these incident reports (IRs) per year which each must be evaluated by experienced plant staff. Of these IRs, only 0.1-0.2% typically are MRFFs. NPPs have found the monitoring of plant SSCs for compliance with the maintenance rule to be expensive and resource intensive.

An alternative approach which has the potential to reduce these costs is to automate the review of IRs through the use of data analytics and machine learning (DA/ML). Advances over the past several decades in computational capabilities and computer programming techniques / algorithms now permit application of DA/ML techniques to analyze the large historical data sets that contain information on the performance of SSCs over years of plant operation. Unfortunately, these techniques have not been widely adopted within the nuclear industry due to its highly regulated nature. A consequence of this structure is that any DA/ML models used in applications subject to regulatory oversight must provide a high degree of confidence in the predictions due to potential regulatory impacts associated with an incorrect determination.

The MRFF Analyzer is a DA/ML application which significantly improves the efficiency of the maintenance rule screening process. The model uses a combination of natural language processing (NLP), Bayesian statistics, and machine learning to identify the likelihood that the incident described by an IR represents a MRFF. The software was trained, tested, and validated on four years of historic incident reports (2015-2019) generated across Exelon Generation's fleet of operating NPPs. This data set consisted of approximately 650,000 IRs. Approximately 420,000 IRs from 2015 through the summer of 2018 were used to train the model, and 80,000 IRs from summer-winter 2018 were used to test the model. Model performance in the field was validated during 2019 with each IR generated by the sites analyzed by the software with review performed by on-site subject matter experts as specified in the plant program implementing procedures. Once the performance of the MRFF Analyzer software was verified as adequate to meet the requirements of the Maintenance Rule, the parallel use of review by human experts was discontinued.

This paper presents the development, training, and deployment of the MRFF Analyzer software across Exelon Generation's fleet of operating NPPs. The algorithms in use in the software are discussed in detail. The challenges dealing with highly skewed data are presented and the approaches used to address these challenges discussed. The results of a detailed sensitivity assessment are presented and used to estimate the expected statistical performance of the software. The overall performance of MRFF Analyzer since its deployment at Exelon Generation is discussed and estimates of the cost savings are provided.

617819 Search: A Symptom-Based Expert for Advanced Response to Chemical Hazards

Abstract:

Chemical exposure accidents pose a risk of enormous personal injury and property damage if initial response is not appropriate. In Korea, from 2014 to October 2019, 1,000 chemical accidents occurred, and 20 people died and 444 were injured. In addition, the number of chemical accidents shows a trend of increasing from 233 in 2014 to 277 in 2018. Existing chemical substance detection technology is focused on the development of response technology for leaks at known sites, and research on the detection at dynamically changing environment and knowledge service for hazardous chemicals considering symptoms is insufficient. Therefore, there is a need for research on a real-time hazardous chemical detection and diagnosis system and knowledge service that protects workers or exposed people from unexpected chemical accidents in the field, by detecting exposed chemicals quickly and helping to cope with dangerous situations early.

In this study, we develop an AI-based system for chemical identification from actively collected exposure symptoms, for real-time response to hazardous material incidents. In order to analyze the symptoms of chemical exposure accidents or contact with substances, information related to chemical substances (symptoms, toxicity, structural information, etc.) is collected and extracted from WISER, PubChem, MSDS, etc. to systemize the related information. The knowledge base of 1,000 core substances is built by designing ontology for symptom knowledge of accident-involving chemical substances. In case of WISER developed by NIH, chemical substances are estimated by simply matching symptom words manually inputted, but the proposed system actively collects the symptoms and predicts the candidate chemical substances (with probabilities) through the knowledge base inference considering the exposure situation. The knowledge base inference uses a knowledge graph as knowledge base that is easy to search, and uses a reasoning method such as the dynamic RDFS++ reasoner, which is easy to infer dynamic data provided by AllegroGraph, and OWL2RL materializer, which is easy to infer static data. In the version 1 of SEARCH, the dynamic RDFS++ reasoner is mainly used to proceed with inference.

Using Machine Learning (k-NN, Decision Tree, SVM, etc.) and Deep Learning, it also predicts toxicity from chemical substance information (structure, etc.) or symptoms from toxicity [1,2] for building expanded symptom KB. The accuracy of the chemical symptoms estimated from the structure and toxicity of the chemical substance is derived from the model with good performance among the previously used models, developing a system with an accuracy of 80% or higher that can estimate 43 symptoms for 1,000 chemicals. Through the combination of two network models that predict toxicity from the structure of chemical substances or predict symptoms from toxicity, it is possible to provide symptom knowledge to be used for symptom detection even for new chemical substances.

SEARCH provides a wide range of information on hazardous substances, including substance identification support, physicochemical characteristics, human health information, and containment and suppression advice. The prototype system is serviced with AI speakers as the main interface and demonstrated at the Gyeonggido Disaster and Safety Headquarters, and plans to combine the established knowledge base and chemical substance-symptom estimation model. The voice recognition interface using AI speakers enables effectively acquire information by saving time and effort rather than directly inputting and matching information manually.

617983 Application of Machine Learning Algorithm for the Analysis of OSHA Accident Database to Identify Trends in Performance Indicators of Process Safety Incidents

Abstract:

Many workplace injuries and fatalities occur in facilities dealing with hazardous chemicals due to unexpected leaks, exposure to dangerous process conditions and other loss of containment events. The U.S. Occupational Safety and Health Agency (OSHA)'s Process Safety Management (PSM) program has been in effect for almost three decades in the United States, requiring companies to establish comprehensive management programs to deal with such hazardous chemicals with the target to assure safety and health in the workplace. Analysis of the trend of process safety incidents in these industries can be a good performance indicator for the effectiveness of the PSM program.

For a long time, OSHA has been maintaining and publishing accident data collected from various industries since the 70's and currently has recorded information of more than 130,000 incidents involving fatalities and injuries arising from various causes. Analysis of such a large database to identify process safety incidents manually is extremely difficult and time consuming. Current study focuses on the development of a machine learning algorithm to screen the database and identify process safety incidents through use of the keywords and incident descriptions associated with each accident provided by OSHA.

In order to validate the machine learning process, the large OSHA database is filtered to analyze incidents from 1984 to current time and only chemical manufacturing and petroleum industries are considered through use of industry classification codes. More than 2,200 incidents are screened out in this manner and are analyzed to determine those that were process safety related incidents. The analysis was done both manually and through use of the developed machine learning algorithm to validate the capability of the latter. The keywords are classified into various categories to aid the algorithm in the learning process. This is the beginning step to the application of artificial intelligence to analyze big databases such as the OSHA database to determine common patterns behind injury and fatal incidents and provide industry with performance indicators that will provide an understanding of where the current laggings are and enable proactive measure to be taken to prevent future incidents.

617448 Game Based Learning – Making the Message Memorable

Abstract:

A perennial frustration for organizations working within high hazard industries is that despite going to great time, cost, and trouble to communicate important process safety messages across the entire workforce, incidents continue to happen. Making these messages "stick" is essential. This presentation will outline a novel approach to developing process safety knowledge and competence through Game Based Learning (GBL).

Based on research that suggests people remember 90% of what they 'do', GBL takes your message out of PowerPoint and broadcasts it interactively to the trainee. A GBL session typically lasts 30-60 minutes, is hands-on and, by using every-day analogies for technical subjects and grounding them in the real world, it is accessible to all personnel across an organization. Indeed, GBL for major accident awareness has been run for school children as part of Science, Technology, Engineering and Mathematics (STEM) teaching for years. GBL comes into its own in high hazard industries where, generally speaking, improvement is about small margins. This presentation will include some examples of games developed by Risktec, and cover a wide range of common process safety themes, to help prompt the audience's ability to replicate similar ideas in their own workplace.

SPRING21
+17TH GCPS
A Joint AIChE and CCPS Meeting

Process Safety Cards – A Good Deal Safer

David Hatch
Process Safety Integrity
Derby, UK
david.hatch@psintegrity.com

Prepared for Presentation at
American Institute of Chemical Engineers
2021 Spring Meeting and 17th Global Congress on Process Safety
Virtual
April 18 - 22, 2021

Process Safety Cards – A Good Deal Safer

David Hatch
Process Safety Integrity
Derby, UK
david.hatch@psintegrity.com

Keywords: Process Safety Cards, Bowties, Gamification, Microlearning, Workforce Engagement.

Abstract

To improve the awareness of less technical personnel to potential Loss of Containment (LoC) scenarios and increase their vigilance to such events & their potential causes so they can be promptly addressed; memorable images are shown on familiar playing cards representing typical LoC events.

1 Introduction

Process Safety Cards have a simple goal – to improve the awareness of a broad range of personnel to potential major accident scenarios and increase their vigilance to such events & their potential causes so they can be promptly addressed. These memorable images represent typical LoC events to deliver knowledge literally into the hands of those who might contribute to and/or be affected by Process Safety Incidents.

With an ageing workforce and increased outsourcing of maintenance and operations activities, it is vital that time and knowledge poor stakeholders (who typically fall outside corporate process safety training strategies) are engaged and equipped to ensure they play their part in sustaining the integrity of assets.

2 Challenge

Process Safety Incidents, Major Accidents or Loss of Primary Containment Events continue to occur in the 21st Century despite better technology and bigger data. For example, on 28th January 2021, a liquid nitrogen release at the Foundation Food facility in Gainesville, US resulted in 6 fatalities and multiple injuries.

A review (by the author) of the 100+ CSB Investigations [1] indicates 200+ fatalities occurred of which approximately 40% were Contractors. This does <u>**not**</u> suggest or imply that the Contractors were at fault, however it does highlight that 3rd parties are often involved in major accidents.

Although Process Safety initiatives, typically referencing Process Safety Management frameworks including OSHA [2], CCPS [3] or the Energy Institute [4], are now commonplace, these may exclude external resources including outsourced Operations & Maintenance personnel who are therefore not as equipped as their internal counterparts to recognize and address potential Process Safety incidents.

We must also acknowledge that wisdom is being lost with the ageing workforce and that assets are getting older whilst those responsible for them are getting younger.

Are we possible focusing on the wrong people and/or protection measures?

How do we put knowledge into the right hands?

3 Workforce Engagement

In order to gain and sustain the attention and participation of the workforce, which may range from novices who need to be informed to veterans who need to be reminded, it is becoming more common to use Gamification methods. These employ game design elements e.g., point scoring, competition with others, rules of play to improve user engagement, learning and knowledge retention.

However, many games require considerable time to setup and play or often cannot be played in the workplace due to the cost and complexity of deploying electronic devices e.g., laptops or tablets to host the games safely (flammable atmosphere ignition potential).

Process Safety Cards offer an alternative approach which offers the benefits of gamification without the time and cost burdens and is based on Microlearning e.g., small learning units absorbed in small time units.

Players are encouraged and helped to learn and apply simple principles:

- REMEMBER Events,
- RECOGNISE Causes (Threats) of Events,
- RESPECT Controls (Barriers) which prevent or mitigate Events.
- REPORT Concerns about the presence or performance of the Controls/Barriers,

Thereafter there is an obligation on the duty-holder or owner/operator to:

- RESPOND to Feedback (concerns, questions, or suggestions).
- RECOVER Protection to its original state.

Using this simple & memorable 'R' model aims to embed these principles into the minds & memories of frontline personnel. It is critical that the RESPOND & RECOVER steps are performed otherwise there can be no faith that contributions are respected and that improvements are being made.

4 Deployment

Process Safety Cards represent 52 different LoC events that occur in the Process industries which are presented as familiar playing cards split into 4 separate suits.

- DEGRADATION Failure within design conditions or "under-strength"
- DISCHARGE Open route to atmosphere
- DAMAGE External impact
- DEVIATION Failure beyond design conditions or "over-stress"

These events are drawn from industry publications including the HSE [5] and CCPS [6] and whilst they are not claimed to be an exhaustive list, they can be attributed to familiar major accidents across a variety of industries.

Cards make it easier to communicate in all directions, top-bottom, side-to-side and inside-outside the organization by providing a common visual language. This gets and keeps the workforce onboard with the Process Safety or Asset Integrity mission.

Sample cards are shown in Figure 1 below:

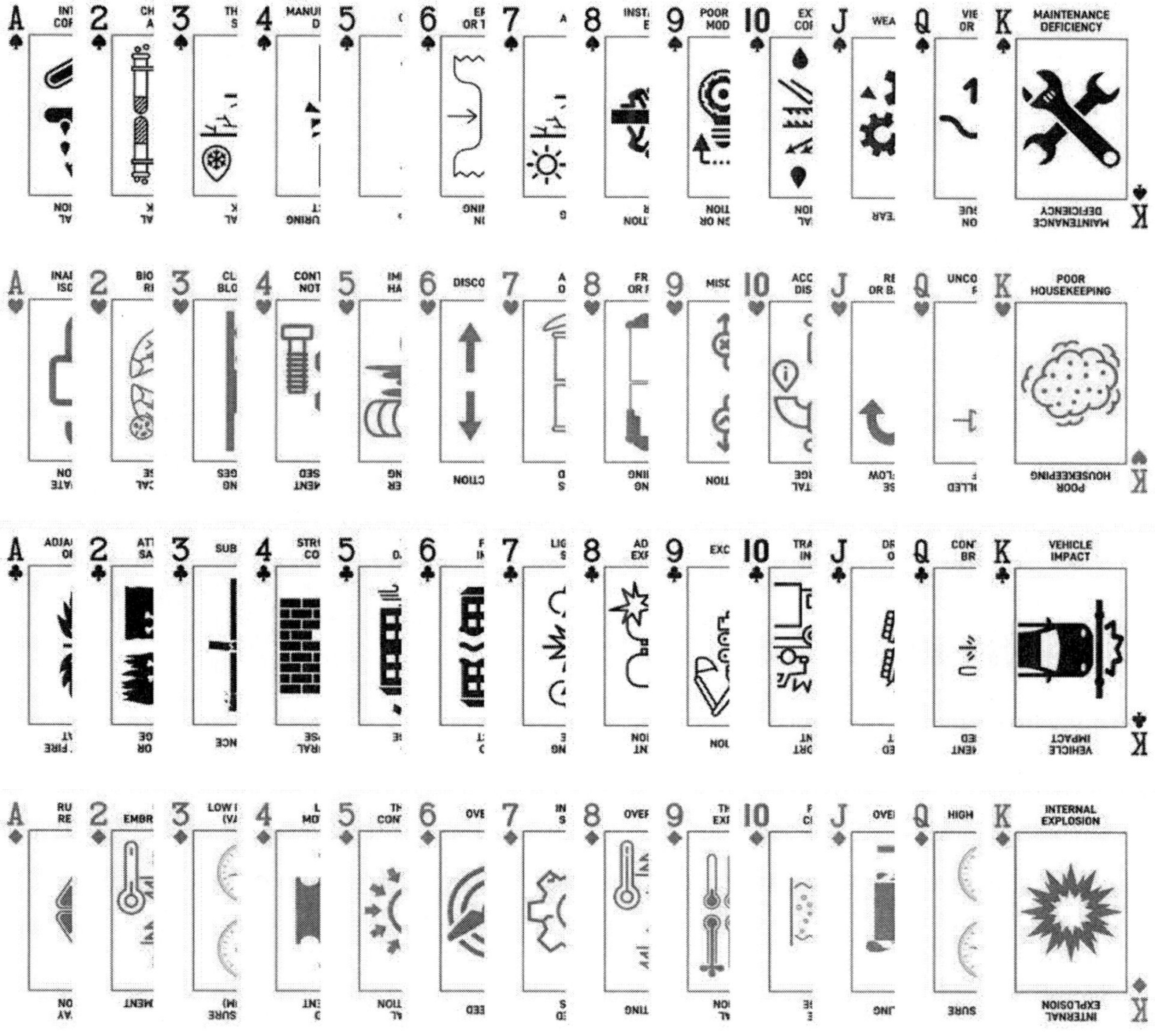

Figure 1 Sample Playing Cards

One of the core concepts of this approach is to "Know your Enemy" [7] where hazards are not the enemy (these are part of doing business) – it is Loss of Containment that is everyone's opponent.

There are several different opportunities to deploy the cards in operational arenas, because they are easy to carry & exchange:

- Simple passive learning (by osmosis) via familiar, informal card games.
- Toolbox talks, e.g., explain your experience of the card you've been dealt.
- 'Pop' or quick quiz e.g., name the event on the card (without title)
- Hazard spotting or surveys e.g., punch hole in cards and tag/tie them onto the plant at vulnerable locations

- Hazard reminder e.g., attach relevant cards to Permit to Work.
- Inductions e.g., visitors, suppliers or short-term contractors during outage, turnaround etc.

The cards are not limited to frontline personnel as they can also be used to engage engineering or management personnel.

- Safety moments e.g., discuss your role related to the card you've been dealt.
- PHA/HAZOP validation e.g., ensure all deviations (cards) have been discussed.
- Performance dashboards e.g., incidents categorized according to LoC type (or card)
- Hazard potential e.g., sum (add) the values of each card i.e., Ace = 11, King/Queen/Jack = 10 & other cards numbered
- Campaigns e.g., each week showcase specific events or cards.
- Relate cards to client or industry incidents e.g., CSB investigations [1].

Just as a pack of cards can create a variety of different games e.g., poker, blackjack, bridge etc. Process Safety Cards utilize the same playing cards elements (or 'pips') which offers an organic approach to gamification that can grow with peer or community feedback i.e., facilities, organizations or industries can develop their own games based on their technology and culture.

5 Scenario Summaries

Playing cards increase awareness of the potential LoC events, however in order to address these, an appropriate level of insight is necessary. Visualizing the LoC scenarios in bowtie format provides a clear, concise summary which is more easily understood and retained than textual or numeric data.

A one-page representation of each type of LoC event i.e., playing card, provides typical:

- Cause (Threat)
- Controls (Barrier)
- Effectiveness or Vulnerabilities (by barrier type)
- Degradation factors (by barrier type)

These are not intended to replace existing analyses and should not be considered sufficient to properly monitor & manage barriers, however they do provide memorable context with a particular focus on the Degradation Factors which create or enlarge the holes in the Swiss Cheese model and are controlled via asset and/or safety management systems e.g., competence, information, maintenance change etc.

A sample is provided in Figure 2 below:

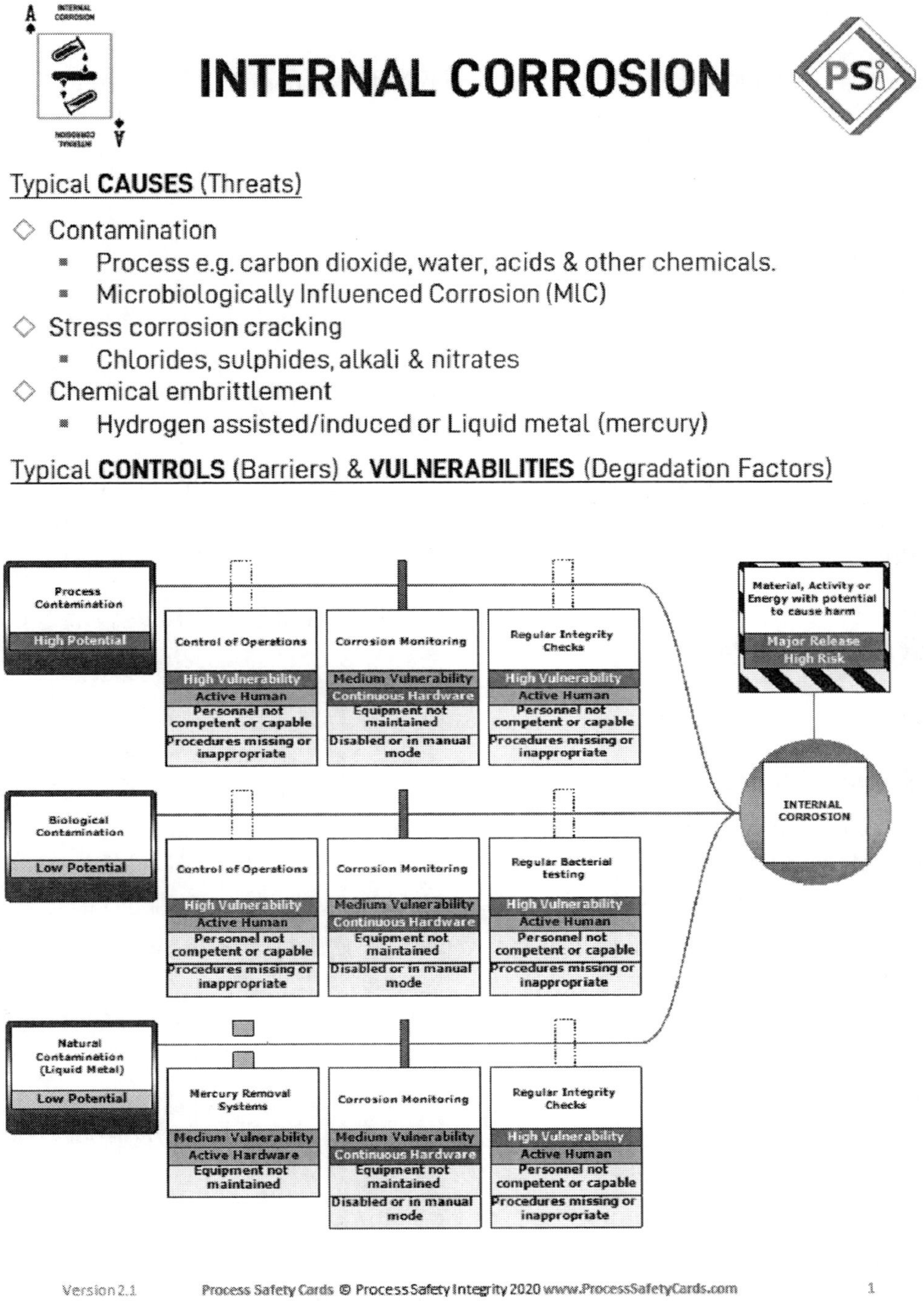

Figure 2 Sample Scenario Summary

Scenario summaries can be posted inside or outside populated or high traffic areas such as control rooms, canteens, workshops, or meeting rooms and rotated on a campaign basis e.g., a different LoC per week. They can also be published online e.g., intranet, SharePoint, or newsletter.

6 Online Collaboration

Although Process Safety Cards are designed to be physically handled, the current COVID-19 pandemic prevents or discourages such personal interaction, therefore digital solutions can be used to use or play the cards via collaborative platforms e.g., virtual whiteboards to connect remote or isolated team members.

7 Conclusions

Although based on gamification principles, the Process Safety Cards are not a game in themselves i.e., there are no rigid rules to follow. They simply provide a knowledge exchange toolkit to facilitate communications (in both directions) between management and frontline personnel to share learning using universal images and local languages.

These are not a new strategy (although they are a novel idea), they simply help deliver or reinforce established Process Safety (Asset Integrity) initiatives which are designed to 'Protect the Protection' i.e., look after the systems (hardware & humans) that look after you.

The cards increase awareness, the bowties enhance understanding, knowing your enemy improves vigilance which drives protection and asset integrity.

8 References

[1] U.S. Chemical Safety Board (CSB) Investigations (https://www.csb.gov/investigations/).

[2] U.S. Department of Labor, Occupational Safety & Health Administration (OSHA) Process Safety Management of Highly Hazardous Chemicals (29 CFR 1910.119) (https://www.osha.gov/process-safety-management).

[3] American Institute of Chemical Engineers (AIChE), Center for Chemical Process Safety (CCPS) Risk Based Process Safety (RBPS) Framework (https://www.aiche.org/ccps/resources/publications/summaries/summary-guidelines-risk-based-process-safety).

[4] Energy Institute High Level Framework for Process Safety Management (https://publishing.energyinst.org/topics/process-safety/leadership/high-level-framework-for-process-safety-management).

[5] UK Health & Safety Laboratory "Loss of Containment Incident Analysis" HSL/2003/07.

[6] American Institute of Chemical Engineers (AIChE), Center for Chemical Process Safety (CCPS) "Guidelines for Chemical Process Quantitative Risk Analysis".

[7] Sun Tze "The Art of War".

618386 Are Your Safety System Barriers Still Work for You?

Abstract:

Process Hazard Analysis is the tool to assess the systems in terms of the process safety scenarios and to determine the ALARP level of the risks that may arise. All the barriers in the Process Hazard Analysis is assumed to be working in the normal operations. However, how you manage to sustain the ALARP level? Do you know any deficiencies and by-pass during the operation? How do they effect the ALARP level? Are you still in the safe side?

All the barriers like Emergency Shut Down Systems, Pressure Safety Valves, Fire and Gas Detection Systems work for your system and you need to follow up in terms of any deficiencies or by-passes made on these. This process starts with reliability of these systems and continues to follow up during the operation. Emergency Safety Systems by-pass are undesired activities but during the operation there would be a need to do this. Here the important thing is how you manage the system with these by-pass and how you follow up these by-pass procedures. Industry and automation systems are improving and there are lots of ways to simultaneously cover all these changes and do the risk analysis to operate the systems safely during these by-pass times. So many companies would be using different ways of controls, in our refinery we have improved a system to follow up all the emergency safety systems with a hierarchy of control and approval automatically including the "Plan B" risk assessments for safe operation of this time periods.The important point here is the type of by-pass (conditional or non-conditional), duration of by-pass (approval hierarchy), what else do you have to operate safely, what other parameters should be followed up by the operator during the by-pass and does he know it?

These automated systems should be covered and controlled during the manual operation or by-pass time intervals in order to eliminate the human errors and not to have any kind of process related incidents. We do need to remember important Process Safety Incidents related with the by-pass of emergency shut down systems like Chernobyl. This solution and automated system would be a barrier not to have such incidents.

Effective response to a power loss event is critical to the safety and mechanical integrity of a manufacturing plant. Power failures can lead to accidents, injuries, loss of primary containment (LOPC), government recordable events, mechanical failure and loss of production. Due to the unpredictability of a power event, there is no single solution or response plan that can mitigate all of the potential issues that a power event creates.

A power event can cause alarm showers/floods which prevent the operator from seeing and reacting to high consequence alarms that could prevent a serious incident if responded to in a timely manner. Based on this, we performed investigation into the last several power failure events at our sites with a focus on the distributed control system (DCS) and its impact on response to power events. This presentation covers the findings from this investigation which include when power events occur, alarm counts/metrics during the events, the severity of each different power event, and findings during root cause investigations (RCI).

We then discuss the different tools that have been implemented into the DCS specifically to aid in operator response to a power event. Some of the tools that are covered in this presentation are DCS graphics "overview" screens for power failure events which provide one-click capability to layer 3/4 process graphics, alarm hiding/suppression concepts, bulk alarm acknowledgement/unlatching techniques, alarm silencing, and automated actions in logic/programming to build more resiliency against an event.

617938 Using Technology to Improve Safety

Abstract:

As the world grows and changes, our business processes evolve, and we must change to meet our new needs. Recent technologies like smart phones, tablets, computers, smart televisions, smart homes, and applications that can manage every aspect of daily life are now widely available in any employee's personal life. These devices and programs reduce the time needed to do routine tasks, providing new opportunities to focus on things the individual may feel are more important or fun. In some cases, these processes learn from previous transactions and tasks and require less of the individual's input as it continues to operate.

However, in a person's professional life, it is common to have a significant portion of one's day composed of filling out paperwork and performing hands-on tasks. Although work instructions, procedures, and policies for everyday tasks are now available electronically, a field employee needs access to a computer to view, access and print documents. This limits the productivity and effectiveness of employees since it forces them to access all information at a single access point or having them carry around paperwork that can easily be damaged by work conditions and the environment. One limitation is unexpected equipment failure. In such cases, it is rare that employee would have all technical information readily available to help troubleshoot in the field immediately, except for minimal visual HazCom labels and warnings. This decreases overall process efficiency by inadvertently creating restraints – specifically, having to go back and collect all required paperwork-on the employee which can be overcome with everyday technology.

Understanding how so much information is instantly available outside of work is important to replicate that level of access and ease in the workplace. At Rain Carbon Inc., our mission is to be "a progressive industrial company committed to collaborative science-driven innovations that result in safe, reliable and efficient operations to ensure outstanding results for our customers, partners, employees and investors." With this as our guide, a new system was implemented to target several areas: safety, reliability, information availability and efficiency.

At our Candiac advanced materials site in Quebec, Canada, integrative software was implemented in 2017 as a pathway to achieving the company's mission. This software provides instant access to process information through phones, tablets and computers, including work instructions, policies, procedures, ability to complete round check sheets and work permits, action assignment and tracking, picture and video capability, report unsafe conditions and incidents, equipment malfunctions, and the management of workforce skills. The capabilities provided by the software have proven to be most effective when used on mobile devices as it allows for information to be accessed directly from any location within the plant helping us improve in all the below points:

- Enhancing site management by providing a live view of notices, work requests and unsafe conditions to immediately communicate and respond to the situation at hand;
- Increasing employee involvement in safety due to higher visibility and lower response times to address unsafe acts/conditions;
- Reducing redundancies of having to go back and forth between paper and electronic work; and
- Decreasing unplanned downtime by providing all relevant technical information instantaneously enabling employees to address deficiencies without delay.

Rain Carbon has pushed forward with innovation and technology by implementing QR codes for all equipment linked to the integrative software solution. This adds a significant amount of efficiency to work processes on the shop floor. When unplanned work is to be performed, the employee can take any mobile device with the software installed on it and scan the QR code on the respective equipment. This will take the employee to a screen with all necessary information regarding the respective equipment, including the work permit and list of materials needed to perform the work in question. A process that would, in most cases, take a matter of hours is reduced to a few minutes and a couple of taps.

Although reliability and efficiency have increased since the system was implemented at the Candiac facility, the most significant improvement has been observed in safety. Prior to implementing this software, safety walks were completed in an electronic or printed version of a Word document, safety incident notifications would be accomplished through phone calls and emails, and near misses would be recorded in a separate system. The centralization of reporting and the enhanced information availability highly encouraged employees to report all safety items with the backing of a more engaged management group, thus enhancing employee safety. Reporting increased with the introduction of the program by 772% on the first year and by 2,150% on the second year. Similarly, during that timeframe, near misses decreased by 40% and first aid cases by 33% compared with the current year.

Results:

From 2017 to 2019, there was a clear increase in reporting after the implementation of the software. Through management support and commitment, we can see a clear decline from 2019 to 2020 YTD of overall identified items, increasing the quality of work and safety across the site.

Implemented in 2017, the program has significantly improved the safety performance of Rain Carbon's Candiac site by engaging management and ensuring all safety items are properly communicated and addressed with as little downtime as possible.

617740 Facility Siting Revalidation - a New Tool

Abstract:

Facility siting studies require periodic review at specified intervals during the life of a facility. As stated in 29 CFR Part 1910.119 (e) (6), revalidation is required every five years based on the previous full study completion date. While revalidation is a regulatory requirement, little is often known about the process of undertaking the initial study or review and, in some cases, this process can become challenging.

A new and simple revalidation tool is available that helps operators determine if their facility siting study needs a complete redo, a retrofit, an update, or a simple revalidation. This paper introduces and explores this simple revalidation tool, providing examples of how it can be used in a plant environment.

617756 Quantifying the Conditional Dependencies between Layers of Protections in Event Tree Analysis

Abstract:

An approach to include conditional probabilities when calculating probabilities of disaster in quantitative risk analysis is presented.[1] Since not all layers of protection are independent of each other, using conditional probabilities leads to more accurate calculations of probabilities of disaster in a chemical process. Based on event tree analysis, probabilities of disaster are calculated assuming that events are statistically independent (independence assumption), and separately considering that events affect each other (conditional probabilities), in three cases with increasing complexity and number of layers of protection. First, a simple alarm – operator case is studied, then two cases involving a chemical reactor and operator with increasing number of layers of protection are analyzed. For all cases, we considered multiple scenarios to show that results do not depend on the input probabilities of particular events. For the latter two cases, the input probabilities of each event were randomly sampled with Monte Carlo simulations. Probabilities of disaster calculated with the assumption of statistical independence were compared to probabilities of disaster calculated with conditional probabilities. In most scenarios of the three cases, findings show that the probabilities of disaster calculated using the independence assumption are lower than those with conditional probabilities, which can lead to undercounting the layers of protection that are needed. Process engineers need to be aware that assuming that events are statistically independent can lead to the possibility of undercounting of layers of protection needed for a chemical process.

Reference: Mott TC, Kivistik PM, Panorska AK, Cantu DC. Comparing conditional probabilities and statistical independence in layers of protection analysis. Proc Safety Prog. 2020;e12215. https://doi.org/10.1002/prs.12215

617799 Comparison of Methodologies for Dropped Objects Analysis in Shallow Water Drilling Rigs in the Gulf of Mexico.

Abstract:

The reform of the energy sector undertaken in 2013 by the Government of Mexico to empower foreign investment in the hydrocarbon sector led to the arrival of private companies with new projects and technology to the Gulf of Mexico. The projects managed in this region move continuously from the exploration stage to the production stage. During this stage transition, Offshore Projects require development of safety evaluations for the design of the platforms that were not common in the Mexican market; among these evaluations there is "Dropped Objects Analysis".

The accidental dropping of objects is considered among the main risks on offshore production platforms since its occurrence can affect personnel (injuries and / or deaths) and damage the facilities (causing spills and / or fires). Thus, as part of engineering design development, there is a need to assess the potential risk associated with impact from oscillating objects and loads that could fall during routine and non-routine lifting activities.

The analysis focuses on the possible damage to the area of operation of a given platform, structures, process equipment and subsea pipelines, evaluating the frequency of dropping combined with the type of damage that would result from the impact. However, for Dropped Object Analysis there are different methods that involve complex calculations, including solving integral equations, which for practical purposes can result in a large time investment.

This work addresses the use of a numerical method that allows a complex calculation routine to be carried out towards a solution with simple arithmetic, thus facilitating the evaluation of the frequency and impact of dropped objects on the platform and reducing the solving time.

As an application case, the Dropped Objects Analysis is presented in two offshore platforms for the exploitation and transportation of hydrocarbons installed in southeast Mexico by a private operator. During the study, the frequency and consequence of dropped objects were evaluated to issue protection recommendations in the design of both platforms and were also compared with the use of traditional calculation methods.

617993 CFD-Based Ventilation Analysis - Basis for Conducting Consequence and/or Risk-Based Assessments

Abstract:

The combination of powerful capabilities in the development of Computer-Aided Design (CAD) drawings; i.e., 3D models; and the performance of formidable computer processors confirm the execution of robust Computational Fluid Dynamics - CFD-based simulations in manageable short-duration periods of time; i.e., minutes. Using CFD tools allow considering the fluid-geometry interaction, key factor for identifying specific features of the site section being analyzed. This information is not possible to be foreseen when using screening tools based on experimental data correlated and/or phenomenological methods.

The present paper illustrates a case study that provides examples of how a CFD-based ventilation analysis can provide insight information of areas that require special attention when a consequence or risk-based assessment is to be performed. The identification of regions with air recirculation, stagnant flow areas with low wind speeds, congestion volumes, etc. verify leak locations and orientations to be evaluated when performing advanced dispersion, fire, and explosion analysis.

SYSTEMATIC APPROACH TO PROCESS SAFETY MANAGEMENT ON ROAD LPG TRANSPORTATION

Presenter/Principal Author
Américo Diniz Carvalho Neto
RSE CONSULTING
Brazil
americodcn@rsem.com.br

Other Author
Adriano Nonato
RSE CONSULTING
Brazil
ext.adriano.nonato@rsem.com.br

Prepared for Presentation at
American Institute of Chemical Engineers
2021 Spring Meeting and 17th Global Congress on Process Safety
Virtual
April 18 - 22, 2021

SYSTEMATIC APPROACH TO PROCESS SAFETY MANAGEMENT ON ROAD LPG TRANSPORTATION

Presenter/Principal Author
Américo Diniz Carvalho Neto
RSE CONSULTING
Brazil
americodcn@rsem.com.br

Other Author(s)
Adriano Nonato
RSE CONSULTING

Keywords: LPG transportation Riks; PSM in LPG distribution, LPG Transport

Abstract

Process Safety Management is systematically adopted in sectors of Brazilian industry, such as oil and gas and petrochemicals, with clear operating safety gains, based on performance indicators. In common, they all use fixed structures and are siting centered to carry out their operations. Also important in the energy chain in Brazil, bulk LPG transportation supplies industrial chain, and business and residences as well. With all this importance requirements for robust process safety management are still missing in LPG distribution sector.
This work aims to implement a safety management system within a company currently responsible for the largest share of the bulk LPG market, considering the particularities of this type of operation, totally dependent on road transport for LPG distribution. To achieve this the first step was identifying the main activities that represent the bulk system's operation, involving bobtail trucks, as follows: supply through road routes; unloading, which is the supply to the customer and the operation of the bobtail truck itself. The strategy of implementing the safety management system considered sequential application of the following steps: 1) Reassessment of the bobtail truck project, aligning it with the latest technological solutions; 2) Conducting a hazard identification study for supply and unloading activities, proposing a customized hazard identification model for

supply; 3) barriers auditing to evaluate unloading components reliability; 4) Auditing operating activities based on CCPS RBPS in order to identify existing gaps. The fifth and final step is the definition of the Process Safety Management approach aligned with road transportation best practices worldwide and in compliance with legal and corporate requirements.

The studies already developed demonstrate that a careful analysis of the main normative reference identified gaps in the bobtail project, with emphasis on passive safety systems, absent in the original project. In the supply risk assessment stage (routes assessment), a specific model for route risk analysis was developed using a customized Excel spreadsheet for risk analysis through predefined menus. This model reduced the assessment time required and the level of subjectivity, allowing the company to develop hundreds of analyzes with the use of video conferencing tools, essential in this case, due to its nationwide scope.

This is a work in progress and the next steps, when completed, will make the company achieve a degree of excellence in Process Safety Management through monitoring for the implementation of all RBPS elements that have been identified as absent or deficient. Many other industrial sectors that carry out transport of dangerous products can benefit from this methodology for implementing RBPS, through the use of the described models, adapting them to the specific modal.

Introduction

It is recognized that chemical transportation risk management is more than just a service; but rather is a wide-ranging system of activities that involves numerous stakeholders with varying backgrounds and interests. To effectively manage transportation risk the entire supply chain, spanning the delivery and handling of raw materials through the offloading and unpacking of finished products, needs to be included and evaluated. In addition to safety and business continuity, the evolving threat of security concerns are now included as part of a comprehensive transportation risk management program. Given the complexity and global reach of many transportation networks, as well as the tradeoffs between safety and security, the management of transportation risk can challenge even the best business plans and programs.

Considering all these aspects and risks involved in LPG transportation, RSE Consulting developed a PSM project with one of the biggest gas distribution companies in Brazil, o order to develop a diagnostic, and create a PSM system to be implemented in the company.

1 Bulk LPG transportation risk assessment

This stage aimed to evaluate a pilot project of adequacy of LPG transport vehicle. The use of common risk analysis methodologies and verification of compliance with international benchmarks were applied.

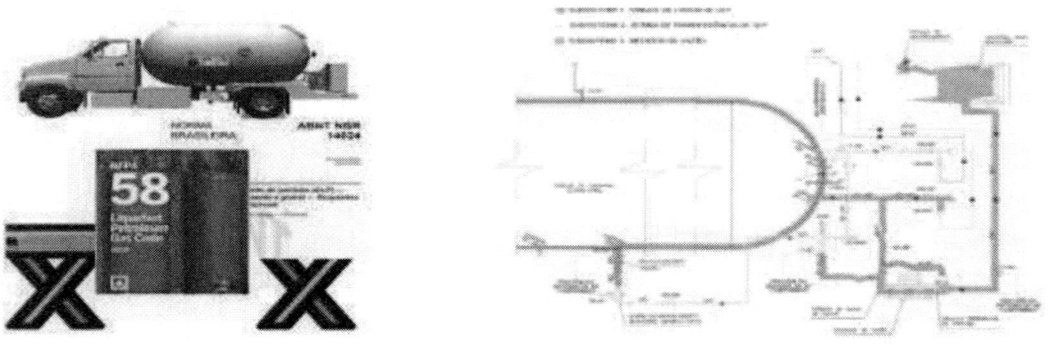

A HAZOP of the system was carried out to identify operational weaknesses

A revised project was obtained considering improvements in management and safety elements, some highlighted below:

Include the identified critical components into the inspection and maintenance plan (Bobtail). Ensure that inspection and maintenance plans have effective tests and tasks for identified and potential damage mechanisms.

Conduct training with operators on the new operating demands (Bobtail), reinforcing the concept of critical components, defining the points that should be evaluated during inspection routines, in order to ensure that the trailers transporting LPG meet the appropriate conservation conditions in relation to the Preventive Barrier of Periodic Inspection of Trailers.

Ensure that truck fleet will have a supply pump interlock system when the minimum level of 5 % of the volume of the LPG tank is reached, to avoid cavitation of the pump.

Every operational unit across the country should comply with managerial based actions concerning standardized execution of maintenance and inspection of critical components.

Include as managerial based actions the adoption of indicators for assets of integrity and reliability as stated on API 754

Define a maximum and minimum speed range to be adopted during the hose rewinding activity on the reel in order to ensure operation compatible with operator safety.

Ensure that a relief valve is installed on the swivel joint. This valve has the function of relieving pressure due to LPG expansion that can occur between two shutoff valves. Ensure that its construction and pressure class meet NFPA 58 requirements

Evaluate the feasibility of deploying a portable remote device for activate emergency activation system.

Modify emergency activation system so that the entire system is shut down or otherwise enforced to fail safe state (pump, pneumatic valves and engine and truck electrical system).

Reevaluate the removal of LPG liquid supply line check valve, which, as designed, will be replaced by a pneumatic-driven shutoff valve.

2. EVALUATION OF TRANSPORTATION ROUTE

The company under assessment supplies LPG for the vast majority of brazilian states

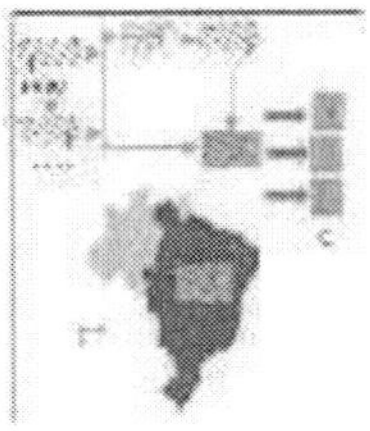

Process safety management on this company is still on its infancy ant still being circumscribed to the staff on Brazilian southwest headquarter. The staff engaged on conducting the evaluation had to overcome the lack of any previous knowledgeabout Risk Analysis of the operating team further hampered by mobility limitations imposed by the pandemic.

A method was developed for route risk analysis as recommended by the CCPS.

All relevant attributes that impacted the risk of the routes were identified and assigned weights for them.

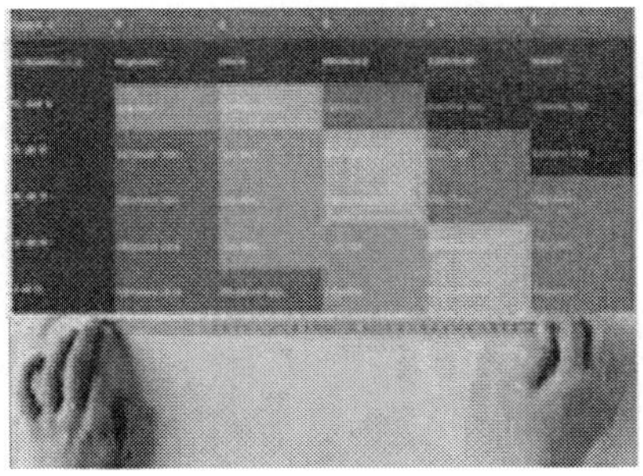

The weights have been calibrated to ensure consistency with the normative risk matrix.

An Excel spreadsheet was developed to assess risk, deploying the afore mentioned risk attributes, allowing teams to engage with maximum objectivity and time savings. Another advantage of the spreadsheet was its smooth learning curve, enabling the use for different teams of the company and possibility of future improvements to evaluate management elements

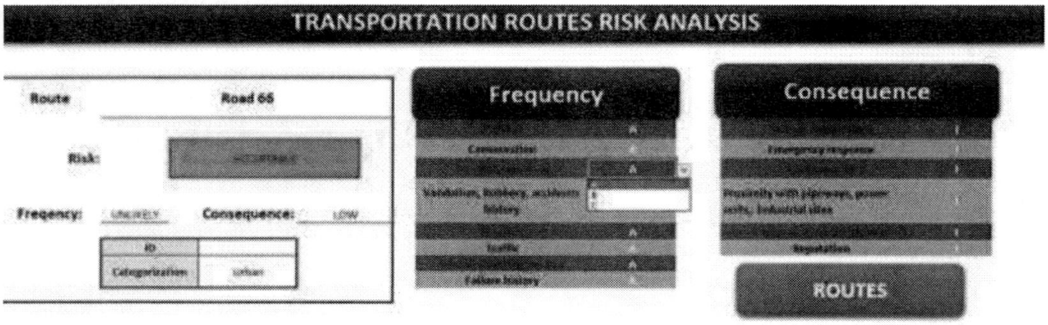

A total of 22 routes were analyzed. It is expected that teams can perform a much higher number of analyses with a minimum of tutoring, given the smooth learning curve for using the tool. Improvement points "appear" from the analysis of the result graphs, allowing recommendations to be properly addressed, as follow:

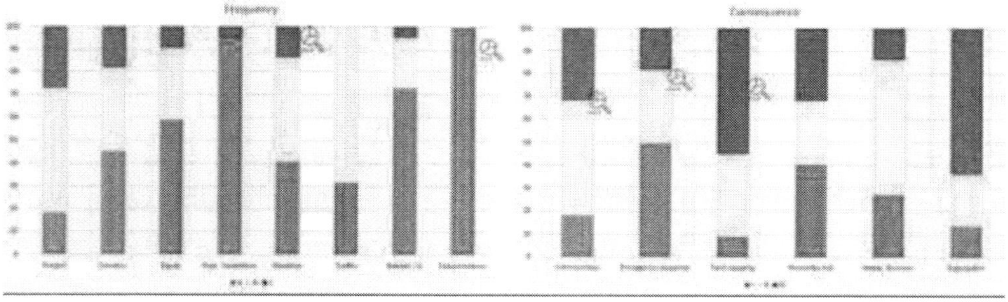

3. CUSTOMER RISK ANALYSIS

The 3rd stage of the project (in progress) will be the application of a bow tie diagram to the customer facilities.

The bow tie will facilitate the process of identifying barriers for subsequent analysis.

4. BARRIER AUDIT

Step to be performed as follows below:

- Definition of critical risk scenarios for the preparation of protocol d andaudit;
- Preparation of audit protocols of critical scenarios and preventive and mitigating barriers;
- Audit of barriers / critical controls of the scenarios defined in the 1st step
- The systems to be analyzed are those of customer facilities. CCPS Bow Tie Approach was the chosen method.

5. IMPLEMENTATION OF PSM

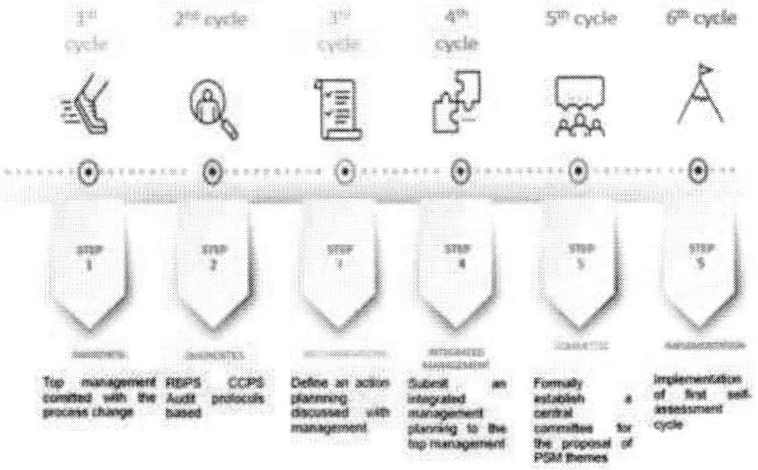

The PSM system implementation strategy is structured in the following steps:

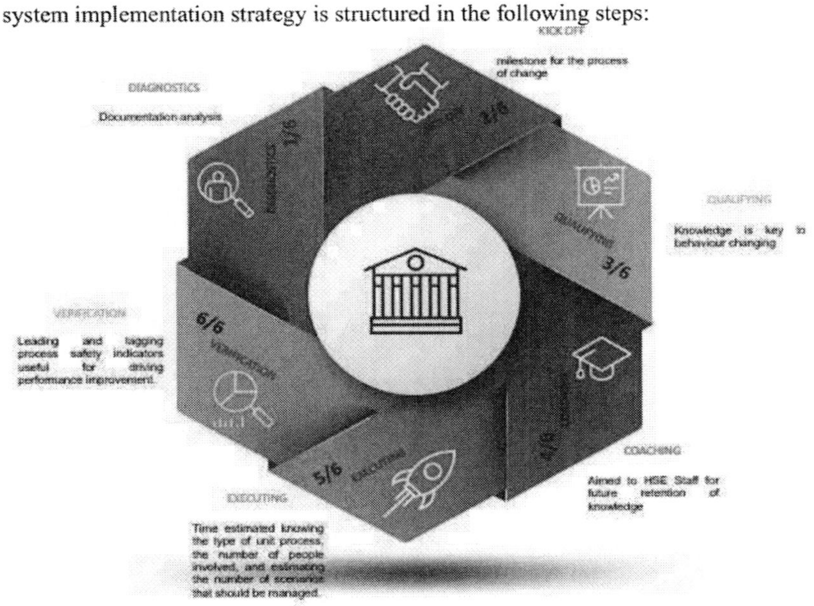

Conclusion

The area of transport of dangerous products, specifically the transport of LPG in large cities, presents many positive contributions to society, however this logistical process has many risk scenarios with high severity in all stages, from loading to the process bases, distribution of the LPG via tank trucks, and refueling at customers' different facilities. This Project allows the organization to not only make a technical diagnosis of the risks of the entire process, but also to set up a Risk Management System that, through the implementation of the requirements of the Management System and procedures, will be allowed to achieve a high performance with the elimination of leakages. gas with the potential for catastrophic accidents.

2 References (examples given to show format).

[1] J.A. Klein and B.K. Vaughen. "Implement an Operational Discipline Program to Improve Process Plant Safety," June 2011.

[2] CCPS. *Guidelines for Chemical Transportation, Safety, Security and Risk Management.* Center for Chemical Process Safety, American Institute of Chemical Engineers, New York, NY, 1995.

[3] RSE. RSE Consulting procedure for Risk transportation qualitative and quantitative risk analysis, 2019. Developed by Professor Americo Diniz Carvalho Neto.

618466 The Evolution of Emergency & Crisis Management in Petronas

Abstract:

Managing emergency and crisis effectively is a paramount aspect in Process Safety Management, especially in the complexity of oil and gas business. Optimum readiness is aimed to efficiently address potential impacts arising from incidents; immaterial of size; to contain the extent of exposure as quickly as possible, with the primary intent of saving lives and protecting the environment. Often we pay too much attention on the left hand-side of the bowtie (control barriers) that we often overlook the right hand-side of the bowtie (recovery barriers). The objective of this paper is to share the evolution of emergency and crisis management underwent by PETRONAS, with the ultimate aim to ensure our interests (people, environment and asset), and reputation are protected.

PETRONAS operates globally in more than 50 countries, hence effective establishment and implementation of emergency and crisis plans are vital. Historically, emergency and crisis management and preparedness has been a reactive science. We had our own share of incidents, and multiple lessons learnt had been consolidated to further improve our systems and barriers. The integration of emergency and crisis management as part of the HSE Management System (HSE MS) and Mandatory Control Framework (MCF) built a strong foundation and governance, and ensures continual improvement, taking into account all the internal and external lessons learnt. Conducting and monitoring the number of emergency drills and exercises, specifically based on the Major Accident Hazard (MAH), is imperative to ensure effective response in case of real emergency and crisis. Learning from real cases raised the vital need of coordinated response and putting the rights interventions and controls in place to minimize the impact of incident. Effective emergency and crisis management is crucial to support process safety implementation throughout the asset life-cycle covering design, operation and maintenance stage.

Throughout the years, we are also continuously enhancing our capability in terms of resources and people. In a nutshell, this paper will talk about the evolution of emergency and crisis management in PETRONAS, where it started with system establishment and capability building, whilst moving towards digitalization to better manage emergency and crisis. This paper will share some real incident cases and how we learn from these incidents to further intensify our emergency and crisis management.

617018 Pressure Relief System Force Balance Assessment – a Review of Its Effectiveness

Abstract:

Design documentation requirements for spring-loaded pressure relief systems are very detailed, based on the definition in API Standard 521, 6th edition. The Installation Documentation section of API Standard 521 recommends provision of inlet and outlet piping pressure drop calculations as part of a complete analysis.

Inlet pressure drop is an important factor, since it is considered that excessive inlet pressure drop will affect pressure relief valve performance, and potentially result in chatter which may damage the valve and piping. In situations where inlet pressure drop exceeds 3% of set pressure for spring loaded pressure relief valves, previous editions of API Standard 520 Part 2 have allowed for an "engineering analysis" to be conducted for these systems - however the content of the engineering analysis has never been well defined.

The latest version of API Standard 520 Part II provides a more detailed description of an engineering analysis and allows a Force Balance Assessment to be conducted for spring loaded pressure relief valves exceeding 3% inlet pressure drop. This method can predict valve stability during relief flow, and identify whether chatter, flutter or stable flow is likely to occur. Valve characteristics such as valve opening and closing times are critical input parameters for this analysis. If stable operation can be demonstrated, then expensive mitigation efforts, such as modifications to piping or the valve itself, can be avoided.

This paper presents an overview of the force balance assessment in a step-by-step manner. Additionally, a sample of spring-loaded pressure relief systems which exceed the 3% inlet pressure drop criteria is reviewed. The aim of this review is to determine the effectiveness of conducting a force balance assessment, and to identify which systems show stable operation, and which systems were identified as likely to chatter.

618390 Chemicals Management inside a Refinery - Does Everyone Knows the Hazards of the Chemicals on Site?

Abstract:

Chemicals are widely used in Oil&Gas industries. Each of them has different phsyical and chemical properties and these differences make it hard to store and use them in a proper way.

In the refineries, those widely used chemicals require a good storage and transportation standards. In order to create a safe environment for those areas, Tüpraş has created its own "Chemical Interaction Matrix". In this matrix, various aspects of chemicals are handled and each Unit has its own matrix.

As a starting point, each of the chemicals are checked as per their stability and reactivity properties from their safety data sheets. The chemicals that do not have detailed information are reacted to each other in the laboratory in order to see if there is a dangerous reaction.

Chemicals that may cause a dangerous reaction to people and environment are stored seperately and the properties of each are shown in the matrixes. These matrixes are hanged on the unit boards so that everyone can see and check if the storage is provided correctly.

It is important to keep these matrixes up-to-date. The matrix is updated when a new chemical arrives in the unit and everybody is informed about update. The update and information is achieved wit the actions that we tracked via our MoC (Management of Change) system.

617653 Operational Risks Involved on Integrating Flare Gas to Fuel System

Abstract:

Flaring is a safe industrial process to dispose waste or excess gas from a plant. But, it has limitations in energy conservation and environmental loss. Flare gas integration to cogeneration plant benefits from utilizing high calorific value waste gases as supplemental fuel to cogeneration plant or gas turbines. Earlier studies on flare gas integration have shown environmental incentive and economic viability. However, the impact of flare gas quality on the operation of cogeneration plants are yet to be identified.

This study delves into the operational hazards and risks arising from the variability of flare gas stream. A novel framework is proposed which can be applied to new or integrated processes which has inadequate information on its operational performance and associated hazards. The study identifies the technical hazards through design data collection, process simulation and sensitivity analysis. Followed by a systematic approach to evaluate the critical operational event occurrences and their impacts through scenario development and quantitative risk assessment.

An important outcome from this study is the identification of critical flare-fuel gas parameters affecting the cogeneration operation through process simulation. High hydrogen content and rich fuel-air ratio in the boiler can affect the boiler operating conditions. Increase in the hydrogen content in fuel system can increase the boiler gas temperature, combustion mixture and flame stability inside the firebox. Risk analysis through Bayesian Network showed a significant risk escalation. With a presumed frequency of 12 hrs/year of high and medium hydrogen gas flaring, there is a substantial rise in the occurrence of boiler gas temperature exceeding design limit and rich fuel mixture formation in the firebox. These top events has major influence on the operational time as well as loss control events - flame impingement and tube rupture.

618388 Process Hazard Analysis Methodologies and Refinery Applications

Abstract:

Process Safety Management System starts with managing the process related risks and from the first step of establishing a new process unit that is needed to evaluate the risks and define the necessary measurements. Process Hazard Analysis methodologies are the systematic tools which help to evaluate the risks in a process. For different needs there are several methodologies like DOW FETI, HAZOP, What if, HAZID, Fault Tree and Event Tree Analysis, Scenario Modelling, Facility Siting etc. Some of these methodologies are semi quantitative some are quantitative methodologies. In order to define and prevent major industrial accidents in our refinery we use DOW FETI, HAZOP, FTA and ETA and Scenario modelling. Moreover for some kind of projects we may also apply What if and HAZID analysis instead of HAZOP depending on the complexity. Facility Siting is another tool for risk related studies which is used to evaluate the risks for buildings in order to protect people inside the buildings.

The workflow for a full unit PHA study starts with DOW FETI calculation which is used for risk ranking of al static equipment inside the unit. DOW FETI calculation is done in accordance with the equipment temperature, pressure, hold-up amounts, components and also reaction types if there is any. Then risk assessment process continues with HAZOP studies which is for scenario identification. In HAZOP, first step is chopping the unit into small parts to be evaluated which is called nodes. After that all nodes are evaluated using P&IDs and PFDs by keywords like temperature, pressure, flow, level etc. HAZOP helps to understand what kind of scenarios are possible inside the unit and these critical scenarios are taken through FTA and ETA studies. HAZOP is a semi quantitative risk assessment method where FTA and ETA are quantitative methods. FTA is applied for critical scenarios in order to find out the probability of top event then ETA is then applied in order to find out the probability of scenarios like fire, explosion, toxic dispersion etc. caused by this top event. FTA and ETA analysis all use the statistical PFD and failure rate values of all initiating events and barriers defined in HAZOP for the related scenario.

Scenario Modelling is the following step which is used for understanding the effect of the probable scenario. The most probable scenarios coming from ETA is modelled to see the effect zones. These effect zones are used to plan Emergency Response Studies for all these potential major accidents.

Facility Siting is another check point in a unit PHA study all potential scenario modelling results are used to evaluate the risks for the buildings and also if there are any in the effect zones of fire, explosion or toxic release then there should be actions defined.

619040 Identification of the Process Hazards By Way of Technical Integration:

Abstract:

Undisclosed threats can lead to **death** and **devastation** in the process industry. Right and constructive hazard recognition supports security procedures, employment health and risk control techniques. It thereby is essential to the health and security of all company employees and to effective business practices.

Without understanding of process hazards, risk cannot be controlled and risk assessment is compulsory for incident prevention. Managers want their workers to be able to recognize not only apparent but evolving threats effectively. Risk assessment needs various **strategies**, **skilled** and testing teams to identify risks and create obstacles to mitigation in their operating environment. Different approaches can enhance the detection of hazards to the risk assessment process.

This paper explains the approach used in the implementation of combined **traditional** and **non-conventional**, **qualitative** and **quantitative** approaches, to identify process hazards, risk assessment and risk management. An insight on the non-conventional techniques *(Temporary repairs reduction & effective permit to work system)* will enable conference audience from all sectors of process industry, in developing integrated process hazard identification and evaluation

617089 Keep the Lion inside the Cage

Abstract:

Keep the Lion inside the Cage

All Chemicals, that we utilize to produce our final Products, are considered as Lions. We have to maintain their healthiness in good condition & under control inside the Plant (Cage) elements (Pipe, Equipment, Vessel, Tank). Process Safety Engineer's Career is an adventure where all safeguards, Barriers & mitigations shall be designed, constructed & operated safely in order to achieve the Process Safety ultimate Goal: "Prevent Loss of Containment". When the chemical leaks, the Lion is free to have high Consequences. With integration for all Reliability, Process Safety & Asset Integrity, we can keep the Lion inside the Cage & eliminate all Probability to have any Consequence whatever its Risk Level.

In order to achieve our ultimate goal, several aspects shall be managed & steward-shipped by the own Organization. There are many Process Safety pillars starting from "hazards identification" till removing the Risk or reducing it to insignificant level; including The Process Safety Competency Development for all Organization levels, Process Safety Knowledge Enhancement, System Support, Technical Guidelines, Risk Management, Process Safety Performance Monitoring and end with the Emergency Preparedness & Response Capability.

The beauty of the Process Safety Concept is the alignment with the Emotional Intelligent that we feel during the dealing with Chemicals (Lions). They shall be always contained & maintained inside their containments. In the same manner of having a specific Process Safety Competency Development Plan for All members of Organization to maintain the Chemicals inside the containment, it shall be available for all members in order to have the Lions inside the Cage. Our ultimate goal is to complete the Journey in safe manner & to maintain the Healthy condition ... in order to keep "Chemicals" inside the pipe & to keep the Lions inside the cage.

617835 Beware of Blind Risk -What You Do Not Know Can Hurt You

Abstract:

This paper discusses the concept of blind risk, how it arises, its dire consequences, and how to best avoid it.

Blind risk is most often due to incomplete, inconsistent, or out-of-date asset and process safety information. Most often, these information quality issues are not obvious or visible, and as such they are a blind risk. They lie dormant, like a beartrap in the undergrowth of a forest until someone inadvertently steps on it. They become gaps in our "layers for protection" -the systems and processes we trust to stop incidents before they happen. But blind risk is also a choice, and it is a measure of a company's tolerance for risk.

Safety, engineering, capital, and operating decisions are only as good as the information they are based on. Most companies may not ***objectively*** know how good (read reliable and trustworthy) the quality of their underlying asset and process safety information is. It does not have to be this way.

There are two things you can do today that will have a significant impact now and into the future:

- Use standardized information quality metrics and tools to measure, monitor and report the true state of asset and process safety information across multiple repositories, applications, and formats. Let all stakeholders including executive management know the good, the bad and the ugly. Use this as a baseline to measure your improvement against.
- Monitor and quantify the cost of poor asset and process safety information. Use this information to justify funding for information quality improvement initiatives.

619136 Relief Systems Work Is a Money Pit: Change My Mind

Abstract:

As a relief system engineer, I revalidate and correct pressure relief system designs daily. Many times, the owner/operator views the pressure relief systems efforts as a 'bottomless money pit' that never returns value and is seemingly never complete. This presentation will dispel this notion and explain how to ensure your pressure relief systems efforts turn into cost-avoidance measures and high-value money-makers.

This three-part presentation will summarize the business purposes of pressure relief system efforts, describe how good relief systems design bases are converted into cost-avoidance measures, and finally share how targeted relief systems efforts can turn into net-positive returns for your facility.

Example content includes safety and regulatory obligations, keeping pace with industry, insurance impact, risk reduction, reputation, safe operating limit changes, and debottlenecking.

ISO 9001-2015 and Risk Mindedness in the Management of Heat Exchangers in Ammonia Synthesis Process

Festus Addo-Yobo, Adrian Lutchman, Sandra Maharaj, Devon Indar
The University of Trinidad and Tobago, Point Lisas Campus, Esperanza Street, Brechin Castle,Trinidad and Tobago.

Introduction

The ammonia synthesis process is one of the most important in the manufacturing Industry because ammonia is a feedstock for diverse products including fertilizer. This process is an interconnected system of production units which include: generation of hydrogen, separation of air to produce nitrogen, ammonia synthesis using catalysts, ammonia condensation and recycling. In the twenty first century efforts were concentrated on the optimizing of the processes either independently or as groups of processes providing inputs to various stages of the processing plant (CEP 2015). As integrated design has been largely successful in optimizing the production plant that is delivered to the producer, attention is now focussed on running the plant so as to optimize its performance. Plant safety, operational efficiency, effectiveness of operation, utilization of resources and productivity are desired outcomes. To achieve these ends, Process Control Engineers must necessarily have conversations with custodians of Quality Management Systems (QMS) in order to optimize the plant performance at a higher management level.

Bayer Technology Services offer Online Optimization & Operational Excellence solutions . Others utilize the TQM and Six Sigma quality assurance models. Nowadays, most plants are run by Distributed Control Systems (DCS) and have modelling software such HYSYS for carrying out such tasks. Thus the opportunity to optimize plant performance online has become feasible. BASF Services have teamed up with Haldor Topsoe to offer Ammonia producers new optimization opportunities online (Process Worldwide 2020).

Ammonia plants have optimized energy recovery through design. Waste heat in the process stream exiting the front end, exiting the secondary reformer, is recovered via heat exchange to produce high pressure steam. Similarly the heat produced by the highly exothermic conversion reaction in the backend is also recovered to produce high pressure steam. The heat exchangers used in these sections (shown in red boxes in the adjacent figure) operate under high temperatures (>400 ºC). They thus are susceptible to failure mechanisms not usually encountered in most petrochemical operations.

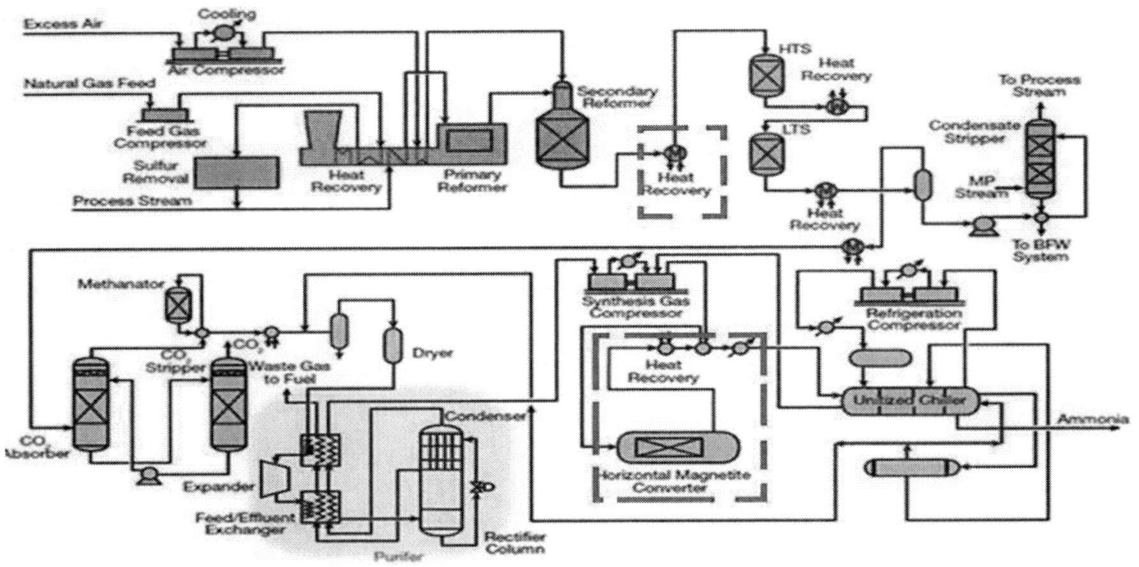

The Quality Management System

An important development in the standard to which quality management systems for production processes must conform is the revision of the ISO 9001 Standard to the 2015 version which makes it a requirement to address risk in all aspects of the QMS and provide anticipatory remedial actions. The standard places the responsibility on top management. Addressing risk is identified in the Standard as having the potential to reduce cost and also providing opportunities. One important way to operationalize risk-mindedness as advocated in the standard is to model the plant performance or the performance of relevant components and investigate scenarios to identify the risks. Heat exchanger tube thinning In the nuclear power generation industry is well researched and reported (Gao et al – 2017, Jonas – 2004, Snachez-Carrera – 1984). Remedial action in the nuclear industry includes scheduled inspection and replacement of heat exchanger tubes prior to their rupture according to protocols using Computer Programmes e,g,CHECWORKS™ (Smith & Horowitz – 2014). In the Ammonia Industry, non-destructive testing can identify tubes that have thinned significantly, which are then blocked off as a preventive measure to avoid water leaking from them to the process stream.

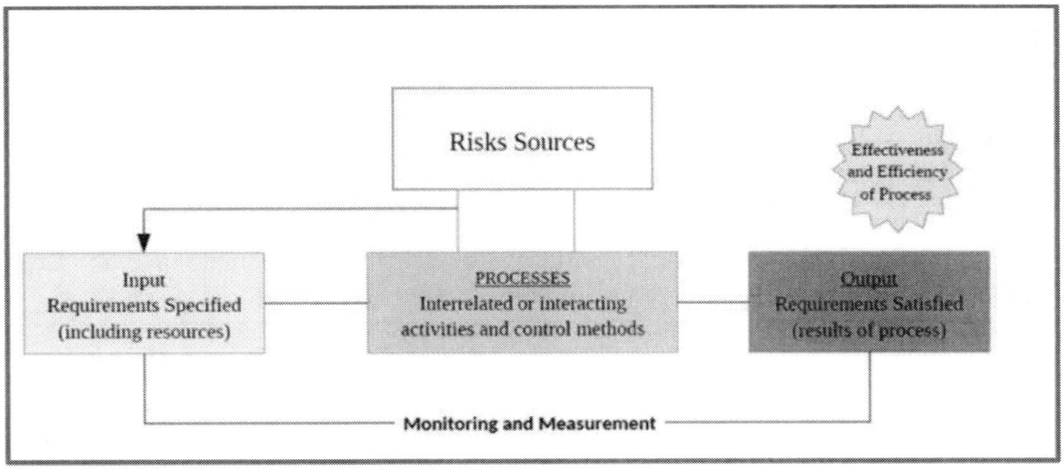

Juxtaposing the great effort and successes in ammonia process optimization to the experience of plant operators, demonstrates the existence of a gap. Even when plants are controlled at desired set points, heat exchanger tubes can and do fail leading to days and weeks of unplanned downtime. Given the high throughput of the plant (600 tonnes/day), this represents significant reduction in plant productivity. Such failures are risks that can be expected, though the events cannot be predicted with accuracy. Heat exchanger networks form an important part of the ammonia manufacturing process. Therefore, these risks should be managed by modelling the operation of the networks. In this work we use HYSYS to model the performance of the heat exchanger network used in typical Ammonia Plant (Chemical Engineering Progress 2015) before and after heat exchanger tubes are blocked off. We investigate numerically, the influence using Boiler Feed Water (BFW) to manage four of the interacting process variables which influence the risk of tube failure: process gas temperature, the velocity of the steam-water mixture in the tubes, the average temperature and the quality of the steam-water mixture in the boiler tubes.

Identification of Risk Factors

The mechanism of platelet erosion has been identified as leading to the thinning of ductile stainless such as the high carbon steel used in the nuclear industry. Abrasive particulates scrape the surface and take off the protective oxide which is swept away by the current of fluid flowing in the tubes, leading to further corrosion. This is known as flow assisted corrosion. Levy (1986) shows that above about 250ºC, temperature speeds up tube thinning process exponentially. Other factors which contribute to Flow Accelerated Corrosion (FAC) are the velocity of the particle, their hardness, the shape and the impinging angle. The geometry of the pipeworks is very important because bends and fittings serve as targets for high velocity dislodged protective oxide.

The HT exchanger tubes which carry BFW for cooling process gases are made of Stainless steel 304L whose mass fraction of chromium are as high as 18-19%. Chromium forms an oxide layer up to 300nm which protects the steel. Should this layer be removed, chromium would react with dissolved oxygen in the feed water to form Chromium Oxide or an even more protective layer comprising a solid solution of Fe, Ni, Cr oxide. Even though the BFW is purified by ion-exchanged to remove chlorides, the possibility of chloride contamination by downstream cooling seawater cannot be ruled out. Chromium can be leached from the protective layer by presence of alkali chloride, which could react with oxygen to release chlorine which further attack chromium oxide and leach out the chromium at temperatures above 500ºC. At lower temperatures, it is postulated that the chloride could react with the oxygen and chromium oxide according the eqn below.

Cr_2O_3 (s) +4KCl(s/l)+$2H_2O$ +1/2 O_2 $\leftrightarrow$ $2K_2CrO_3$ (s/l) + 4HCl.

The chromate formed would either remain in solid phase in the absence of liquid water or dissolve in the hot BFW water and be leached away. In either case, no contiguous layer of protective chromium oxide layer would remain and the iron underneath would rust. This mechanism would lead to thinning. Thus, the risk factors are: the contamination of the BFW with chloride, high temperatures and the oxygen levels in the BFW. The steam weight would also constitute a risk factor. The presence of nitrates is a risk factor as it known to corrode SS304L. These two mechanisms would apply to all boiler tubes in all the heat exchangers and thus all the tubes would be equally exposed to this risk.

In an effort to increase heat transfer coefficients in heat exchangers in the ammonia process, "static" mixers are often inserted in boiler tubes. The increase in the heat transfer coefficients is accompanied by the disruption of the laminar boundary sub-layer which also leads a higher mass transfer coefficient, thus increasing the potential for leaching chromium and iron ions from the metal. The additional risk factor associated with presence of static mixers in tubes is the higher velocities in the pipes they cause by occupying space.

The tube inserts pose additional risks introduced at the design stage. They are pendula bodies which will undergo complex vibrations and translational motions during plant operation. The amplitude of the vibrations will depend on the flexural rigidity of the metal. Their motion pose two additional risks of tube thinning: they can impose forces normal at the tube wall which could disrupt the formation of replacement protective oxide layer (F. Addo-Yobo Unpublished work); they could also make contact with the tube wall, peel off existing protective layer and set off vibrations in the walls which can continually weaken the replacement oxide layer and increase its porosity to migrating metal ions thus accelerating erosion and corrosion thinning of the steel underneath.

Thus the operational risk factors are: steam Weight, temperature in boiler tubes, velocity in the boiler tube. Design risk factors are: presence and motion of inserts for promoting turbulence, vibrations; and Utility risk factors are: contamination of process water with chlorides and nitrates after the ion-exchange stage.

U-tube heat exchanger

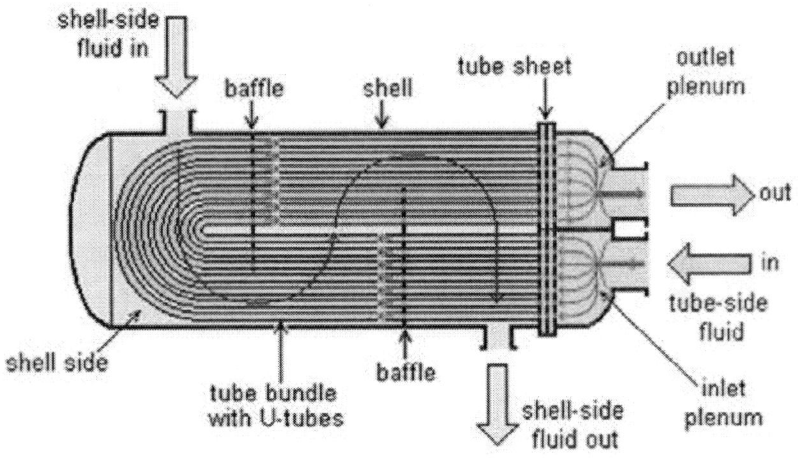

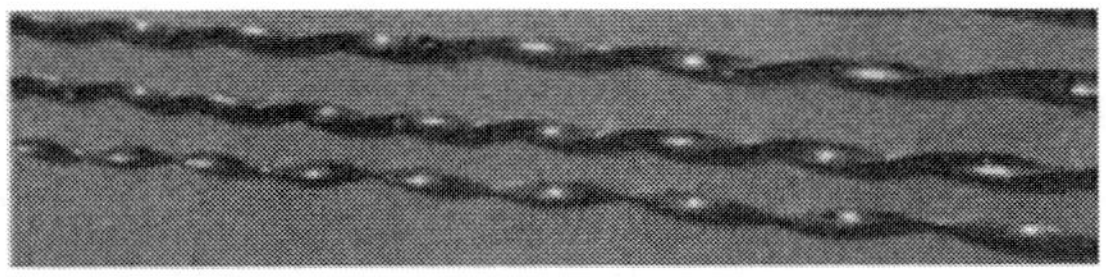

The Models in Hysys

he heat exchangers enclosed in the dotted red boundaries in the Process Flow Diagram above are modelled as a network. E2A is immediately downstream of the ammonia converter, E2B is the second in that stream. E111 is the exchanger in the front end stage of the process. The flow of Boiler Feed Water for cooling, in the boiler tubes in the heat exchanger E2A, E2B and E11, commence as single phase then becomes two phase with the generation of steam as heat is transferred from the process gases on the shell side. The dispersed two phase model in HYSYS was adopted to account for this.Whereas standard HYSYS models for heat exchanger could not accommodate finned and unfinned tubes in one exchanger, the rigorous EDR models were modified successfully to include this capability without changing the hydraulic and heat transfer characteristics. We were also able to incorporate variations in the process gas specific heat with composition without much computational expense.

From heat balance, the weight fraction of steam in the tubes (v_f) relates the process gas flowrates ($\dot{m}_G$) and temperature drop (T^G), BFW flow rates ($\dot{m}_G$) and changes in enthalpies (H^W). The steam weight was computed as shown in the adjacent equation.

$$v_f = \frac{\frac{\dot{m}_G}{\dot{m}_W}\left[C_{pc2}(T_1^G - T_2^G) + C_{pc1}\left(T_{1,in}^G - T_{2,out}^G\right)\right] + H_1^W - H_2^W}{H_2^v - H_2^W}$$

Thus confirming that the flow rate of the BFW is a suitable manipulated variable for controlling gas outlet temperatures. As stated above, the velocity of BFW in the tubes is important and its value can pose a risk. Its value varies with the weight of steam. The dispersed two phase model for steam-water mixes in HYSYS was used for modelling the flow and heat transfer to BFW in the tubes of the heat-exchanger. The shell side heat transfer coefficient in HYSYS are predicted by the well-known Dittus-Boelter equation.

The models built in HYSYS accurately predicted the output pressures and temperatures of the process gas entering and leaving the three heat exchangers, under standard conditions, within 3% percent of the expected values, as shown in the Table1. The simulations were carried out under standard operating conditions at 100% plant rate. The table also shows the risk values of these deviations from the TEMA values, which will be discussed later.

Results and Discussion

Often process Engineers find the need to increase the throughput of a Plant beyond the design values in order to improve productivity. However, this may carry additional risks of shortening the lifespan of the heat exchangers. In order to protect the heat exchangers and avoid overloading them, the product stream can be spilt and a percentage diverted around exchangers E2A and E2B. Table 2 compares output variables when model plant is operated with higher throughputs than the design levels. Columns 4,5,6 of Table 2 below show that the tubeside temperatures can be maintained within 0.3% of the design values with the bypass arrangement. However, steam weight increases beyond the prescribed TEMA values of 30%, thereby increasing potential operational risk. The weight fraction of steam increases beyond prescribed TEMA values with additional feed rate of 7% to 17%. Columns 5 & 6 of the above table show that at around 17% excess feed rate, increasing the cooling water rate by around 8% reduces the steam weight fraction to values close to the prescribed limit, thereby reducing operational risk and at the same time raising production rate.

The studies show that when 5% of tubes in E2A are blocked to mimic the response of industry to tube thinning beyond a certain level, to eliminate the risk spilling water from burst tubes wetting the product gases, the steam weight remains at reasonable levels close to the TEMA limit. However, the velocity in the tubes increase to values close to the prescribed TEMA limit because of the reduced cross sectional area available to the flow of BFW. The risk values of plugging tubes in exchanger E2A on exchangers E2B and E11 are negligible. Tables 3 summarizes the predictions of output variables caused by the plugging 10% of the tubes in E2A on E2A. The simulations show a reduction in the area available for heat transfer, in exchanger E2A by about 3% accompanied by a reduction in the heat duty of ca 4.3%. The reduced cross-sectional area available for flow on the tubeside results in 11% increase in the mean velocity through the tubeside and a 7% increase in the maximum velocity in the tube. The latter is close to the prescribed TEMA limit thereby increasing the operational risk posed to the plant by the blocking the tubes. The effect of this increase in velocity could be a shortening on the service life of E2A.

Table 1: Comparison of the predicted plant output variables with the design historic values.

	Mass Flow Rate	% Deviation of predicted Temperature from Standard TEMA values	Risk Value	% Deviation of predicted Pressure from Standard TEMA values	Risk Value
EE11 Gas Inlet	Standard	0.00	None	-1.85%	None
EE11 Gas Outlet	Standard	-0.21%	None		None
EE11 BFW Inlet	BFW rate	-0.39%	None		None
E2A BFW Inlet	BFW rate	0.18%	None		None
E2A Gas Inlet	Standard	0.00	None	1.31	None
E2A Gas Outlet	Standard	-0.05%	None	2.62	None
To **** (E2A BFW Outlet)	BFW rate	-0.29%	None	-1.33	None
BFW	BFW rate	2.53			None
E2B Gas Outlet	Standard	1.57%	Negligible	2.80	None

Table 2: Effect of Increasing the Plant Throughput

KEY PROCESS PARAMETERS	Design Specification or Historic data	100% PLANT RATE	7% increase in feed process gas	17.4% increase in feed process gas	16.7% increase in feed process gas

Process Gas Rate (kg/s)	Standard	No deviation	+7%	+17.4%	+16.7%
Deviation in Water circulation rate (Wt/Wt)	Standard	No deviation	-17.2%	-21.9%	-15.7%
Deviation in steam weight $\frac{Predicted - Tema\ limit}{TEMA\ limit}$	-23.3%	-23.3%	+33.73%	+46.8%	+13.4%
Risk Status E2A due to steam weight	None	None	Highest	Highest	Highest
Deviation in tube-side outlet temperature of E2A	None	-0.29%	-0.14%	-0.143%	-0.15%
Risk value of tube temperature E2A	None	None	None	None	None
		Mean/Max			
E2A TUBESIDE % Velocity deviation	No Data	-61%/-20%	-64%/-3.7%	-63%/+3.6%	-63%/-3.7%
Risk Value E2A %Vel deviation	N/A	No Risk	/Localized Medium Risk	/High risk	/Medium risk
E2B % dev in TUBESIDE VELOCITY	-51%	-59%/-57%	-66%/-64%	-66%/-64%	-63%/-61%
Risk value	No risk	No risk	No risk	No risk	No risk
E11 TUBESIDE VELOCITY (m/s)	1.92	-60/-58	-66/-64	-66/-63	-64%/-61%
Risk value in E11 tube side velocity		No risk	No risk	No risk	No risk

Table 3: The Impact of blocking off 5% (10%) of the tubes Exchanger E2A of Performance of E2A

Heat Exchanger Characteristics	E2A with no plugged tubes at	E2A with plugged tubes	Percent Deviation

	theoretical conditions	at conditions from DCS	
Effective Surface Area available for Heat Transfer (m^2)	*	*	-3.07
Heat Duty (kW)	*	*	-4.35
Dev from Standard in Mean Temperature of Outlet of tubeside E2A	-0.15%	-0.16%	
Risk value	No Risk	No risk	
Corrected Mean Temperature Difference (°C)	*	*	7.34
Deviation in Heat Transfer Coefficient	*	*	-8.1%
Mean Shellside Velocity (m/s)	-56.2%	-56.1%	0.28
Maximum Shellside Velocity (m/s)	-53.2%	-53.2%	0.00
Risk value of deviation in maximum shell side velocity	No risk	No risk	
E2A TUBESIDE % mean Velocity deviation $\frac{mean\ velocity - Tema\ vel\ limit}{TEMA\ velocity\ limit}$	-63%	-59%	
Risk Value E2A %Vel deviation	None	None	
Dev maximum Tubeside Velocity (%) $\frac{Predicted\ max\ velocity - Tema\ vel\ limit}{TEMA\ velocity\ limit}$	-6.7%	-0.23%	
Risk value of Predicted maximum velocity	Medium risk	Medium risk	

Discussion, Conclusion & Recommendation for Future work

The studies have demonstrated the usefulness of building a rational model of the operation of the heat exchangers in the Ammonia process based on the diagrams presented in CEP (2015). They confirm the remedial actions taken by the Industry to improve productivity and maintain safe operation are sound. However, they point to the unfortunate fact that the quest to increase the heat transfer coefficients in the U-tube heat exchangers by inserting turbulent promoters are fraught with design risks that might not be possible to eliminate.

The simulations coupled with the present understanding of the mechanism of corro-erosion of high chromium stainless steel products (Folkeson et. al. 2011) suggest that once tube thinning is allowed to commence, caused by a risk factor introduced at the design stage, e.g. the inclusion of unfastened tube inserts, which increase heat transfer coefficients but which disturb the protective oxide layer, the exchangers may have to be eventually de-commissioned because of the difficulty of laying down a contiguous protective layer as shown in the schematic shown above right.

In order to firm up on the mechanisms that trigger tube thinning by corro-erosion of stainless steel boiler tubes in the Ammonia Process, further research into the risks posed by tube inserts and the careful weighing up of the cost of mitigating this factor as required by the Standard need to be carried out.

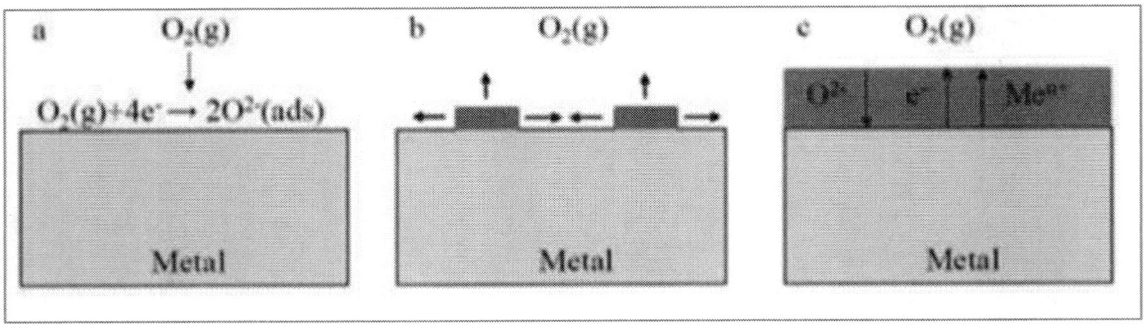

Schematic of the initial formation of an oxide scale by adsorption of oxygen, oxide nucleation and growth to form a continuous oxide scale (stages a–c).

References

1. Venkat Pattabathula & Jim Richardson 2016, Introduction to Ammonia Production, Chemical Engineering Progress pp 69-75

2. Process Worldwide: 12:08:2020 Editor Ahlam Rais. Haldor Topsoe, BASF to Offer Ammonia Producer new Optimisation possibilities. ID 46800904 https://www.process-worldwide.com/haldor-topsoe-basf-to-offer-ammonia-producers-new-optimisation-possibilities-a-955495/

3. The Role of CHECWORKS in an Effective FAC Programme, David Smith and Jeffrey Horrowitz , ICONE 22, 22nd International Conference on Nuclear Engineering, 2014 Prague

4. Gao, Weimin, Li, Yungang and Kong, Lingxue 2017, Numerical investigation of erosion of tube sheet and tubes of a shell and tube heat exchanger, *Computers and chemical engineering*, vol. 96, pp. 115-127, doi: 10.1016/j.compchemeng.2016.10.005.Boiler tube failure in Heat Exchangers for Power generation in the Nuclear Industry

5. L.E Sanchez-Carrera, "Mechanism of Corrosion-Erosion in Steam Extraction lines in Power Station", PhD Thesis Dept of Mechanical Engineering MIT 1984

6. Jonas, O. (2004). FAC and Cavitation: Identification, assessment, monitoring, prevention. *Combined Cycle Journal*.

7. Suez Water Technologies and Solutions. (2020). *Handbook of Industrial Water Treatment Chapter 11- Preboiler & Industrial Boiler Corrosion Control.* Retrieved from https://www.suezwatertechnologies.com/handbook/chapter-11-preboiler-and-boiler-corrosion-control

8. Md. Moniruzzaman Bhuyan1, Ujjwal K. Deb, M. Shahriar, Simul Acherjee 2017, Simulation of Heat Transfer in a Tubular Pipe Using Different Twisted Tape Inserts, Open Journal of Fluid Dynamics, 2017, 7, 397-409, http://www.scirp.org/journal/ojfd

9. Folkeson, N., Jonsson, T. Halvarsson, M. Johansson, L.-G., Svensson J.-E., The influence of small amounts of KCl(s) on the high temperature corrosion of a Fe-2.25Cr-1Mo steel at 400 and 500°C. *Materials and Corrosion*, (2011), 62(7), pp. 606–610

10. F, Addo-Yobo 2021, Motion of Pendula inserts in heat exchanger tubes in the Ammonia Process and their impact on exchanger risk, Unpublished work.

617202 Optimal Placement of Gas Detectors Using Dispersion Modeling

Abstract:

Historically, gas detector placement was performed by experienced designers using rules of thumb and judgment based on fundamental principles of gas cloud behavior. Unfortunately, designs using these approaches have been inconsistent and, in many cases, poorly performing. Historical data from the UK HSE shows that certain fixed gas detection systems routinely were only able to detect less than 70% of "major" gas releases. In the past decade, advances in computer modeling and 3D graphics has enabled the implementation of chemical process quantitative risk analysis (CPQRA) in the service of gas detection system design. The general process for using CPQRA to develop and verify a gas detection performance metric called coverage, which is essentially the fraction of the all release scenarios that are detected by a gas detector array, has been documented in the International Society for Automation (ISA) technical report ISA TR84.00.07 – Guidance on the Evaluation of Fire, Combustible Gas, and Toxic Gas System Effectiveness. This tutorial provides an overview level explanation of the workflow for the process of gas detection mapping, including the risk-based selection of performance targets and quantitative verification of the achievement of the coverage performance targets using dispersion modeling results and leak rate estimations that are similar to the proven approaches used in CPQRA. The tutorial goes through the entire process of gas detector placement for a sample facility where liquefied petroleum gas products are stored, beginning with risk assessment and going all the way through a completed gas detector layout.

Presenter Information	Contact Information
Presenter Name: Bejoy K. Bharatiya	**Contact Name**: Bejoy Bharatiya Telephone No.: +61 413 098 008
Presenter Biography: Bejoy Bharatiya was Process Safety Advisor at DNV GL Offshore and Industrial Services (B) Sdn. Bhd. He has over 30 years of professional experience in the upstream oil & gas, refining, petrochemical, and mineral processing industry. Prior to joining DNV, he held various positions, Process Safety Group Lead at Saudi Aramco Shell Refinery Company (SASEF), Senior Process Safety Engineer at SANTOS (South Australian Oil Search) Pty Ltd. Australia, Senior Operations Engineer with Saudi Aramco Gas Operations, and the Principal Process Engineer of Kellogg, Brown & Root (KBR), Australia.	Fax No.: E-mail Address (Required): **bejoy2014@gmail.com** **pinkubejoy@iinet.net.au** Your Company/Department:

Abstract (max of 1500 words)

HSE Philosophies in Capital Projects of Petroleum Refinery and Petrochemical Complex - A Practical Approach

Bejoy K. Bharatiya

The petroleum industry has long since recognized that certain risks are involved in the major hazard industries. It becomes a major challenge to minimize risk by comprehensive HSE framework in the project execution of a major process facility in the hydrocarbon industry. Over the years, the oil and gas industry has learned lessons from a number of process safety events and various hazards such as loss of control and loss of containment top events that could result in: fires, vapour cloud explosions (VCE), toxic and asphyxiant exposure, which occurred in various parts of the world and are not limited to the country of origin. The design and operation of process facility for the refinery and petrochemical complex have been the subject of major hazard to the oil industry.

It has been the continuous effort of the industry to update the design and operating philosophies to improve the inherent safety of the processing units. The present paper is concerned with defining the major safety considerations of international oil companies, and Industry Standards in designing and operating the refinery and petrochemical process facilities. HSE philosophies are required to guide and drive the design intent of HSE critical systems. The main safety philosophies include: ESD and blowdown, fire and gas detection, toxic gas (H2S) detection, active and passive fire protection (including guidelines for firefighting systems and fire protection), layout HSE philosophy, process safety and containment etc.

This paper provides the basic safety design guidelines for a major project but does not attempt to cover in detail every aspect of safety design. Reference should be made, therefore, to specific discipline philosophies and specifications as appropriate.

The primary environmental philosophies include venting and flaring of hydrocarbons and H2S/SO2, CO2 production, solid waste and effluent wastewater disposal, energy efficiency etc. In the project planning and execution, these philosophies must be effectively embedded in the design through the HSE premises document.

For the major project execution, the HSE philosophies are written in a way to instruct the design functional specifications to be compatible and aligned with the existing facility, but are updated to take into account of new technology, new knowledge and other industry requirements.

HSE philosophy shall be considered throughout all the project development phases, engineering, construction, start-up, and operation via dedicated and well-structured efforts. The safety philosophy, and resulting protection of persons and assets, is combining all provisions undertaking during all project phases.

HSE Philosophy for the detailed engineering phase is prepared with the objective to elaborate the Project design philosophy with respect to identification and elimination of hazards and to promote a design that is inherently safe. Objectives of this HSE philosophy are:

- Provide a safe working environment for personnel
- Minimize the potential for hazardous occurrences
- Avoid exposure of personnel to potential hazard
- Contain and minimize the effects of hazards
- Provide adequate safe access and escape away from hazards
- Protect the environment
- Provide a design that is fit for purpose for field life with the required integrity for safety and environmental protection

HSE Design Philosophies

During the detailed engineering, and subsequent construction and commissioning, particular care will be taken by achieving:

- Good engineering practice
- Application of an all-encompassing QA system
- Development and implementation of a safety management system
- Design reviews at 30%, 60% and 90%, layout review, HAZID, Design Safety Review (DSR), accessibility reviews and other reviews.
- Qualitative and quantitative risk assessments and IPF review
- Design of installation for extreme weather conditions
- Execution of an extensive pre-start up review of the plant before introduction of feed (Hydrocarbon, gas) covering both hardware and, control systems and procedures.
- Conformity check with local regulations, project standards and other applicable industry standards and directives.

Hazard assessment – Hazard assessment is based upon the hazard of the materials present in the processing facility. Some of the highest-ranking typical hazards included in the risk assessment (to be taken from facility Hazard Register) are:

Hazard	Top Event	Consequence
All Liquid Hydrocarbons (HC) including condensed flare gas	Loss of Containment	Fire
Toxics / H2S / N2 /BTX (Aromatics-Benzene, Toluene, Xylene)	Loss of Control or Containment	Personal Exposure
All HC, including H2 enriched HC	Loss of Containment	Generation of vapour cloud leading to VCE or fire
Heavier than air HC	Loss of Containment	Vapour cloud and/or Fire resulting from heaving than air HC

The design for the active/passive fire protection systems, fire detection systems, gas detection systems and general HSE requirements shall comply with the HSE Philosophy, project standards and referenced codes/standards, specifications, and regulations.

The fire safety of a plant is the degree of reduction in vulnerability to fire and/or explosion incidents in terms of their probability of occurrence and associated potential damage. This degree is maximized by the integrated application of fire prevention, fire protection and fire-fighting measures.

The philosophy is to achieve the desired degree of fire safety primarily by fire prevention measures, which will also result in an optimum availability of the facilities to produce. Where despite the preventive measures potential fire hazards remain, fire protection measures are taken. The basic objective of fire protection is to limit or prevent the escalation of a fire to avoid risk to life and to minimize material damage.

The Project shall be divided into separate fire risk areas of possible fire involvement. This separation may include major pipe racks, roadways, or vacant space. For estimating firewater demand, only one major fire occurring in one fire risk area at any time is to be considered. The fire water system shall be developed to meet the highest firewater demand of the worst-case firewater scenario. Extension of the existing firewater system on the Refinery to cover the new installations will be considered. Harmonization of fire safety measures shall be maximized.

In addition to fire protection measures, the plant fire-fighting organization is available to control and/or extinguish fires that occur despite the fire prevention and fire protection. All shift operators to be trained in industrial firefighting form the first intervention team (FIT) and additional fire fighters will be available from the Fire Station Crew. The combination of FIT and the existing Fire Station Crew will form an effective team, available within minutes after an incident is detected. If required, extra support can be provided through the external fire brigade.

There shall always be more emphasis on fire prevention, fire protection and fixed and semi-fixed firefighting systems than on the capacity of any fire-fighting organization.

The existing fire safety facilities in terms of mobile fire-fighting equipment and fire brigade organization of the refinery complex will be used to a maximum extent.

Furthermore, as the refinery hydrotreating and Sulphur Recovery Units process/handle large quantities of H_2S/SO_2, special emphasis shall be made during the design phase for the installation of suitable system for H_2S service such as high mechanical integrity (e.g. materials, pipe types, double pumps seals etc.), safeguarding, H_2S detection and automatic isolation/shutdown system. Design of this system shall be as per project standards and actual HSE philosophy for H_2S detection system.

Safety relief venting shall be provided for equipment, which is subject to internal pressure build-up from an external heat source. Pressure relieving devices shall be designed in accordance with project standards and/or industry standards.

Plant layout and spacing shall be developed based on project standards and industry practices, e.g. API RP/STDs, General Electric GAP Services (GE GAP) 2.5.2 "Oil and Chemical Plant Layout Spacing. Safety distances will be optimized to reduce the area utilized once more detailed design is progressed and risk studies have been completed (e.g. Physical effect modeling).

This paper discusses industry experience and summarizes the salient design considerations and risk assessment to improve safety integrity in the execution of capital projects for refinery and petrochemical complex. An effort has been made to highlight the HSE methodologies adopted by major oil companies using project standards, and industry best practices and standards like API, ASME, NFPA, IEC, ISO etc. Emphasis has been made to increase the awareness among the discipline design engineers and process safety engineers to develop safer designs of the refinery and petrochemical facilities.

619065 Addressing Aging Processes and Equipment

Abstract:

Number of the fertilizers plants across the globe have been installed and are in operation since many decades. The operation with the old equipment and conventional process techniques is being managed coupled with budget constraints. Hence the old aged plant equipment have high risk of failure under the shed of outdated processes, and may cause severe damage to the plant assets as well as to human. Consequently, one has to bear more and repeated maintenance cost and regular replacements of equipment.

Fatima Fert Plant (Previously Daud Hercules) also commissioned in 1969 and is in operation to date. Dealing with the challenge of old aged equipment and process techniques, very strict and robust reliability plans have been established and implemented to inspect the equipment with the advanced ultrasonic techniques and to make sure that the asset is in operation under controlled parameters. Based upon the finding of inspections, 10 year rolling plan have been developed to replace the equipment that are prone to failure.

The paper will be providing the details of the challenges and risks faced due to old age equipment and the way forward to mitigate and control that risk with optimized budget utilization.

619392 Why Is There a Regularity of Process Incidents Occurring, or Is It All Just a Re-Occurrence of the Past?

Abstract:

We attend conferences to hear the experiences of others, learn lessons from past events and take that experience and learning back to our place of work. Yet a simple analysis of suggest there is a plateau in the progress to reducing the frequency of process safety incidents.

So perhaps the question to address is, "Why are lessons not being learnt"?

This paper will highlight various data demonstrating the "plateauing" of performance; highlight particular recurring indicators that need to focused on and look at how learning may be distributed to those that need to increase their knowledge of PSM. There will be a particular emphasis of the use of case studies in this approach.

The paper will particularly interest those still on the learning journey in PSM and those questioning alternative approaches.

618397 How You Manage the Actions Coming through Process Safety Management Systems?

Abstract:

For improving process safety in the plant, we are putting great afford managing task of operators. Alarm management is important for control operator's tanks. Quantity of the alarms should not be more then control operator can manage.

However, there is gap in engineer's tasks management. Workload of an engineer is an important parameter in complex plants. This workload can be rapidly increase with HSE related tasks. As a result of Management of Change, Incident investigations, ... most of the tasks are given to operations engineers, projects engineers, maintenance engineers.

If the workload is too high it is easy for the engineer to miss the important actions. With HSE Action Portal, we aim to give management a view of task load for each department. The department responsible can focus on the overall tasks of the department

HSE Action Portal make possible to reach actions and tasks coming through PSM system elements such as Management of Change, Pre-Start up safety review, incident/accident investigations and process hazard analysis in the same platform. Users are able to analyse their action/tasks in terms of number and completion percent. In this way, users are triggered to follow closely their actions and complete them in a correct way.

It is an useful tool also for top management in order to collect data and create statistics periodically. It can be seen actions/task situations for each refinery and department.

617997 Guidelines for Risk Ranking Hazardous Scenarios When Conducting PHAs

Abstract:

Most of the companies in the process industry use corporate guidelines, or dedicated criteria and principles for conducting Process Hazard Analysis (PHAs). Examples of factors addressed are the following: nodes selection, process deviations to be brainstormed, level of detail when writing causes/consequences, safeguards categorization, risk matrix to be used, and how to risk rank the identified scenarios.

This paper collects different approaches from leading PHAs at several companies. An analysis comparing these approaches is provided and strengths and weaknesses are identified. Finally, based on safeguards and consequences categorization, guidelines are proposed for consistently risk rank hazardous scenarios.

619107 Practical Guide to Operationalize & Visualize Cumulative Process Safety Risks Perspective

Abstract:

The paper describes how Process Safety Management can be seamlessly integrated with Operational Risk Management. It demonstrates how operational processes like Permit to Work, MOC, Job Hazard Analysis, Incident Management and Process Safety Management can be brought together to support day-to-day decision making based on the actual cumulative risk status of the facility.

Key thinking is developed to answer questions such as *How do we make Process Safety Risk as visible as HSE* and *How can we prevent previously identified hazardous sequences of events from reoccurring?* , thus increasing the entire system safety.

This paper presents practical approaches in overcoming the challenges in collecting data from disparate sources and learnings on realising Dynamic Operational Risk Management (ORM) from various client implementations. In addition, it touches upon topics such as *How to analyze the data* and *Use of machine learning to predict future risk* .

The paper presents a risk calculation methodology that calculates barrier status and visualizes cumulative risk:

in a standardized barrier model, based on IOGP 544 standard, in Swiss-Cheese view

geographically by site and area/location using site graphics

per critical scenario in a BowTie view

It connects and visualizes process safety risk by automatically collecting data for Tier-1,2 and 3 safety indicators that are covered by various processes spread across the organization.

Furthermore, it shows how the cumulative risk status can be used in operational processes, potential conflict detection and planning of activities to manage and minimize the actual risk.

Keywords: Barrier Management, Process Safety Management, BowTie Methodology, Swiss-Cheese Model, Operational Risk Management, Project Implementation Methodologies, permit-to-work, MOC, JHA/RA, Machine Learning

ENGINEERING & OPERATIONS

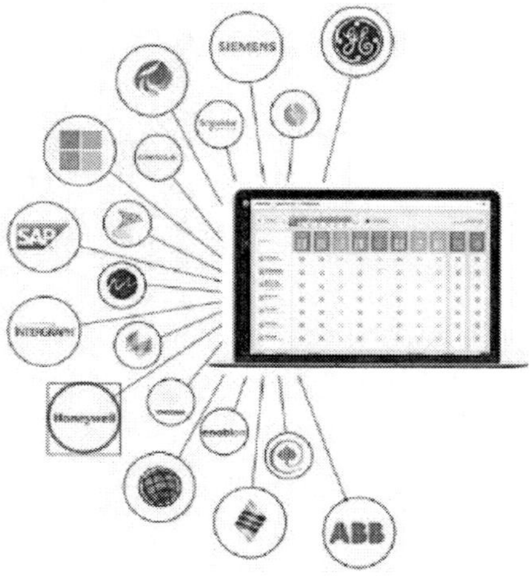

BARRIER MANAGEMENT INTEGRATION DATA POINTS

Integrity Data: CMMS, Audits & Assurance, Competence Management, Inspections, Engineering data.

Operational Intervention: Process Control System / Historian, Shift Logs, Operator Rounds, Permits, Temporary defeats.

Barrier Challenges & Failures: Incident Management, Management of Change, Process upsets, PSM events.

BARRIER MANAGEMENT – PROCESS SAFETY MANAGEMENT

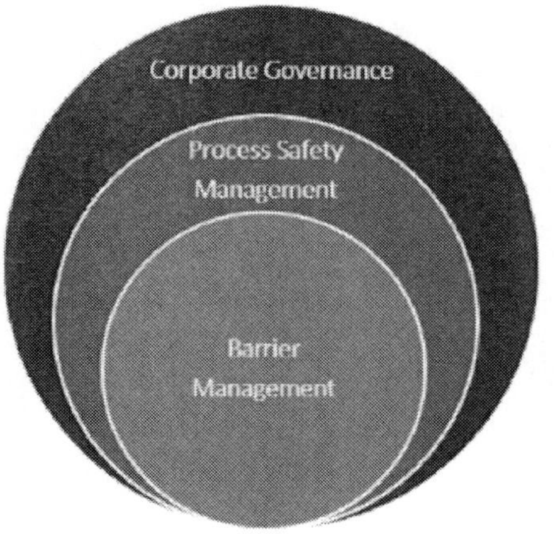

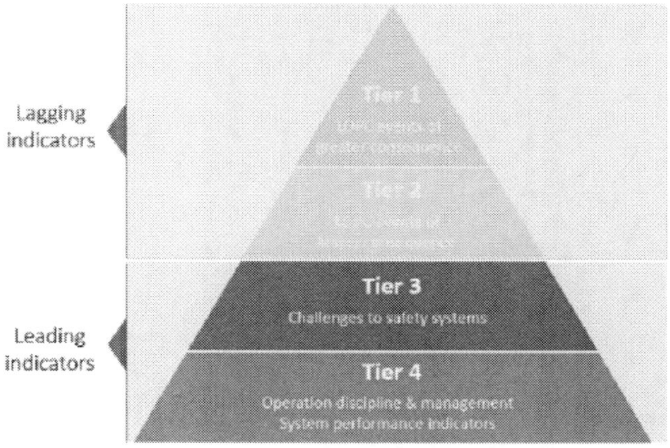

The focus of Barrier Management

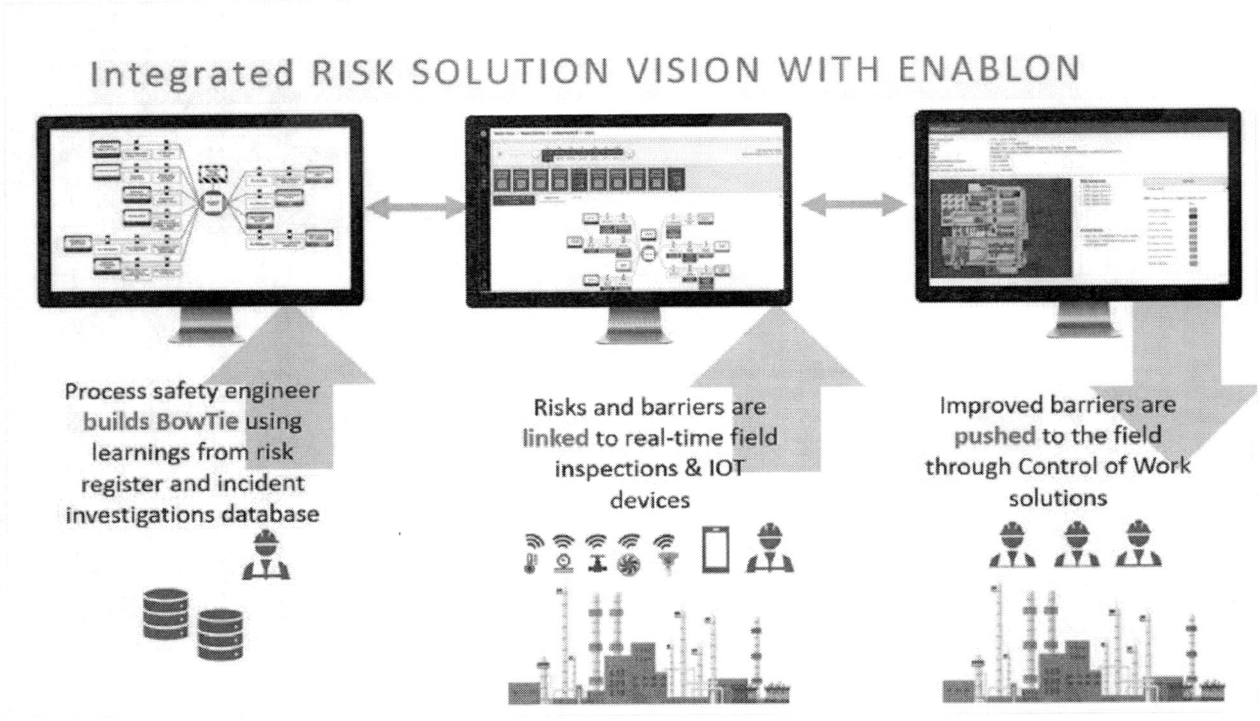
Integrated RISK SOLUTION VISION WITH ENABLON
Process safety engineer builds BowTie using learnings from risk register and incident investigations database
Risks and barriers are linked to real-time field inspections & IOT devices
Improved barriers are pushed to the field through Control of Work solutions

619126 Controlling Human Performance Between Both Unplanned and Planned Tasks within Abnormal Operation Mode

Abstract:

There is confusion in terminology used in the chemical-related industry for the class of procedures commonly referred to as Abnormal Mode of Operation and Abnormal situation management. This paper provides a clear definition of each mode of operation and gives examples of how the human performance is controlled for each.

1. Normal – either a continuous mode or a normal batch mode of operation.
2. Planned Non-routine (Non-Normal)– Startup, shutdown, are online maintenance are the main non-routine or non-normal modes of planned operation. But planned Temporary procedures, with time limits, are also part of these.
3. Abnormal– These include those activities covered by generalize procedures or guides (planned in a general sense) the activities that don't have written, step-by-step procedures (but noting that many companies have guides for most of these abnormal situations). A further breakdown is possible:
    1. Response to upsets using a Trouble-shooting Guides(TSG) for handling deviations from the operating window (these or normally triggered on an alarm); these may ultimately lead to a shutdown, safe park, or emergency shutdown (all of which are also proceduralized), if the deviation cannot be corrected in time.
    2. Response to failures using a Temporary procedure - such as how to run in bypass mode if the flow controller fails and you want to keep running in manual mode.
    3. Response to Unanticipated Events - The FIRST occurrence of this is handled by what is generally referred to as Emergency-MOC, which is really saying we will make a change immediately, and do the risk review of the change later, and then learn from this one case and proceduralize into a Temp Procedure or TSG for the next time it comes up.
4. Emergency Operations- a diminished or reduced operating plan; normally a Temporary procedure (similar to 3.b. above)
5. Emergency Shutdown - a TSG that fails to resolve the issue in time will go to this; or for some events such as a sudden loss of containment, we go straight to these
6. Emergency Response- normally in conjunction with 4 or 5, but focused on the protection of people, assets, environment, given the release or other imminent harm is in play.

This paper will provide a framework to ensure no mode of operation is overlooked and it will help sites understand what is needed to control risk during each mode of operation. Most of the time in the paper and presentation will be focused on classification 3, Abnormal Mode of operation.

619218 Proactive Lopc Indicator Critical Analysis through Power BI Dashboard

Abstract:

This work applies Power BI dashboard in order to proceed a critical analysis of proactive loss of primary containment indicator so that possible process problems can be detected before greater events occur by prognoses and decision-making to direct resource management in a more assertive way.

A process safety event is an unplanned or uncontrolled release of any material including non-toxic and non-flammable materials from a process, or an undesired event or condition that under slightly different circumstances could have resulted in a release of material (API 754). The pyramid model can represent the predictive relationship between lower and higher events.

It is important to guarantee the good register of pyramid base events (Tier 3) to achieve enough information, identify the fail installation's weakness, avoid higher process safety events, reduce losses and improve the facility performance. Nevertheless, the only register is not enough to achieve these improvements. Its's necessary to quantify and analyze the data to guide the decision-making. A critical analysis involves the following parameters: leaked product, fail causes, equipment that failed, engineer discipline, LOPC location, consequences.

The first step to make the critical analysis is having a reliable database. This step involves training the operational team in leading and lagging process safety indicators and process safety management. The communication must be direct and clear. Furthermore, it is necessary to provide a satisfactory registration platform with the necessary fields to carry out the subsequent critical analysis.

After that, it is possible to extract all the information to proceed the proactive LOPC indicator critical analysis. At this point Business Intelligence (BI) presents itself as an adequate tool. Power BI, from Microsoft, is one of the most prominent BI tools. It is a collection of software services, applications and connectors that work together to consolidate, make information from different data sources coherent, visually engaging and interactive (from an Excel spreadsheet to a cloud-based data collection).

Through the LOPC critical analysis using Power BI dashboard it is possible to evaluate the Tier 3 incidents, focused on developing a learning approach. Therefore, it is possible to prevent and monitor failures trends and effectiveness of implemented actions. The critical analysis permits to direct the resource management by identifying the equipment and systems with recurrent failure events. It is worth mentioning the effective contribution for Process Safety Culture, since the leadership, operation and related areas could see the cause-effect relationships between the minor events and the daily operations.

619342 A Case Study on Ensuring "ALARP" Demonstration to Justify Continued Operation with Respect to "Mah" By Integrating Process Safety Techniques for Existing Process Facility

Abstract:

A lot have been written on process safety. However, practical applications of all these theories & studies has its own challenges. At the end, after overcoming the challenges as well as developing, implementing & monitoring strategies, a risk to existing facilities can be reduced to ALARP/ acceptable level.

This paper will throw light & discuss about requirements to integrate different process safety strategies, challenges during implementation & their solutions to demonstrate that existing oil & gas facility is safe for operation, through a practical case study of one complete process train. The methodology includes;

- Identifications of process trains
- Data collection
- Data Review
- Review past changes & their implementation
- Update records & P&IDs
- Carry out/ Update Integrated PHA with process simulation
- Development of Failure Scenario
- Risk Calculation considering existing safeguard (preventive and mitigating measures)
- Risk submission and ALARP demonstration
- Compare risk
- Identification of safety critical elements (SCEs) & Safety Critical Tasks (SCTs)
- Development of Performance Standard of SCEs
- Developing management system of PLAN-DO-CHECK-ACT
- Integrating the same in Audit/Inspection
- Develop training through simulations

Challenges faced by team for existing facility to achieve optimum process safety & ensure ALARP level of risk, and their solutions will be discussed.

Conclusion

This exercise provides evidence that wherever practical, the design and operations of the existing oil & gas facility has attained a level of risk that Justifies Continued Operation with respect to major fire and explosion events. This demonstration of Justification for Continued Operation is based on the number of safety measures inbuilt at design stage, Quantitative Risk Assessment and various other process safety studies revalidated during operational phase of the oil & gas process facility as well as after implementing changes through MoC. It must be noted that a separate process of action tracking is in place to prioritize and monitor implementation of recommendations provided to reduce the risk further to As Low As Reasonably Practicable (ALARP). A systematic approach of Plan-Do-Check-Act is developed to ensure that, safety critical tasks are carried out in timely manner by competent personnel, to ensure safety critical elements are in place and functional in line with performance standard.

616668 Towards Management Systems Perfection – Successes with Implementing a Layer of Verification Analysis to Continually Treat Process Safety Risks

Abstract:

Trust, but verify, by continually checking everything! This paper will demonstrate the importance of verification as a critical step in continually improving the treatment of process safety risks, including implementing inherently safer designs and operating practices, and controlling residual risks.

A well-established design concept for management systems is the continual improvement lifecycle involving the steps: Plan-Do-Check-Act (PDCA). Similar to standards for managing quality, environmental, occupational health and safety, and security, models for effective process safety management (PSM) systems also utilize the PDCA cycle. At a higher level, the PSM elements of Mechanical Integrity, Incident Investigation and Compliance Audits address verification in the "Check" step; while every management process of the PSM System should incorporate a "Check" step with verification processes and procedures to monitor and assure compliance and performance. This paper will present a taxonomy of activities for meticulous verification against policies, objectives, plans and other requirements for a vibrant PSM system, showing the importance of examining, analyzing, assessing, confirming, inspecting, testing, and reviewing planned activities, desired conditions and achieved results.

Attendees will benefit from practical takeaways of how each management system element can be developed with "verification designed in" and a verification culture created to treat process safety risks at three opportune times: before the incident (prevention), during the incident (emergency management) and after the incident (incident investigation, root cause analysis and lessons learned).

A comprehensive case study with practical examples will be presented to demonstrate a proven technique of Layers of Verification Analysis (LOVA) as a best practice integrated into the conduct of a process hazard analysis (PHA). The case study will show the audience "how to" enhance the PHA facilitation wearing an "auditor's hat" to raise questions and verify the effectiveness of existing safeguards, all of which are managed by the PSM System. Borrowing from the basic principles of management systems auditing, LOVA asks these questions for each hazard scenario: are systems, processes and procedures associated with causes (of human errors and equipment malfunctions), with consequences (of physical phenomena and resulting impacts), and with safeguards (for both engineering and administrative controls) *(1) in place?, (2) adequately designed? and (3) effectively implemented?* LOVA has also been used for supplemental analyses within the PHA to address human factors, previous incidents, facility siting and changes (e.g. MOCs). Recommendations from the LOVA in PHAs can be used as lessons learned for improving the implementation of the PSM system on a daily basis, without having to wait on the next periodic audit or PHA revalidation. The LOVA technique has been shown to be equally suitable for any stage of the process lifecycle, including inherently safer design evaluations.

The paper will also present three incident case histories to show how missing, inadequately designed or ineffectively implemented verification processes and activities have contributed to process safety incidents.

This paper will draw upon the authors' extensive experiences in designing, developing, implementing and auditing PSM systems, as well as conducting PHAs, building siting evaluations, and incident investigations.

617112 Study and Experimental Validation of CO2 Blowdown at Cryogenic Conditions

Abstract:

CO_2 solidification is one of the critical issues being identified in cryogenic technology process, particularly during the blowdown process. This poses critical challenge in developing the technology for separation of CO_2 from natural gas as there are limited studies and data to support the design and simulation of the process. Depressurization of high CO_2 fluid at cryogenic temperature would lead to possibility of CO2 solid formation, hence potential blockage of process equipment and venting line.

The new cryogenic technologies are currently being tested for the proof of concept. Hence, a pilot plant, which is a scaled down version of the technology was developed. One of the major challenges faced during the pilot plant testing is the emergency depressurization philosophy as the process involves CO_2 solids handling which is uncommon to the industry standard. Lack of experimental data of CO_2 blowdown at these conditions presented a technical gap which needs to be addressed to properly predict the solidification and mitigations, hence avoiding over-design of the plant and chances of accidents and unscheduled downtimes can be minimized.

Therefore, this paper will focus on the experimental study which have been done on CO_2 blowdown scenarios at a pilot plant. The pressure and temperature profile provide real insight of the solidification of CO2 during several blowdown scenarios. Finally the paper would validate a simulation data conducted using the experimental data as one of the potential prediction tools for CO2 blowdown process.

617588 The Supervisors Next Big Challenge: How to Ensure Operational Discipline on the Work Fronts

Abstract:

The Supervisors Next Big Challenge:

How to Ensure Operational Discipline on the Work Fronts

There is a problem that the petrochemical and oil and gas industries have not solved effectively: how to transform work fronts in environments based on operational discipline.

All the commitment and effort that an organization mobilizes in its risk management, finally, is updated in what happens between supervisors and operators on the work fronts. That is why it is so necessary to work on the alignment of the workforce of operating companies and contractors.

In the last decade, Senior Management became more aware of its role: defining what were tolerable risks and assuring resources. Process Safety, Integrity and Environment ceased to be discursive themes, to effectively have a genuine place at the small table where strategic decisions are made. The biggest challenge has been getting it to operate according to those definitions.

Supervision is carried out on the work front, the true executive arm of the organizations in the field. And for this organizational segment to ensure a disciplined operation, they need to be able to manage the culture.

A common mistake in the risk management approach has been to think of the lack of operational discipline of supervision as a problem of isolated individuals, and not of the context in which these people work, that is, of the context that the activities provide. organizations, from the context produced by culture.

When analyzing the supervision segment, one should not think of undisciplined people, but cultures that tolerate them, or that reinforce risky behaviors and shortcuts to the established norm. If supervision does not ensure operational discipline on the work fronts, it is because the organization does not provide the conditions for this to occur. People are the reflection of the culture in which they operate. Individuals do not impose a personal style of managing risks in isolation, but adapt to group agreements, displaying safe or unsafe behaviors depending on what is collectively tolerated. Field supervisors only take risks that the culture allows them to take.

The change of mindset in the supervision segment should not be built on the basis of shortcuts or "crash plans", but rather in avoiding in a progressive and lasting way the repetition of incidents, through a transformative process: generating a culture of discipline, learning and mutual trust in the work fronts, based on group agreements between operators and contractor companies, managed by a trained supervision to make decisions based on risks, combat the normalization of deviance and keep in mind the sense of vulnerability.

The key is to generate the conditions for effective learning. In addressing group problems, and not individual ones. The challenge is to create the conditions so that operational discipline is not a declaration of principles but a reality in the field. For this they do not need new knowledge, but to deepen what they already know, in the right context.

The pandemic, paradoxically, provided an opportunity to improve the quality of supervision development, personalizing it. Although online self-training has been around for a long time, it is not the most appropriate tool to generate changes in operations. For that, culture needs to intervene.

The next decade seems to keep a central challenge: transforming supervisory teams into custodians of culture and guarantors of risk management, through work fronts based on operational discipline.

Presenter Information	Contact Information
Presenter Name: Bejoy K. Bharatiya	**Contact Name**: Bejoy Bharatiya Telephone No.: +61 413 098 008
Presenter Biography: Bejoy Bharatiya was Process Safety Advisor at DNV GL Offshore and Industrial Services (B) Sdn. Bhd. He has over 30 years of professional experience in the upstream oil & gas, refining, petrochemical, and mineral processing industry. Prior to joining DNV, he held various positions, Process Safety Group Lead at Saudi Aramco Shell Refinery Company (SASEF), Senior Process Safety Engineer at SANTOS (South Australian Oil Search) Pty Ltd. Australia, Senior Operations Engineer with Saudi Aramco Gas Operations, and the Principal Process Engineer of Kellogg, Brown & Root (KBR), Australia.	Fax No.: E-mail Address (Required): **bejoyb@hotmail.com** **bejoy2014@gmail.com** Your Company/Department:

Abstract (max of 1500 words)

Project Assurance Review and Audit Framework for Capital Project of a Petroleum Refinery - A Process Safety Approach

Bejoy K. Bharatiya

The petroleum industry has long since recognized that certain risks are involved in the major hazard industries. It becomes a major challenge to minimize risk by **comprehensive project assurance/audit framework** in the project execution of a major process facility in the hydrocarbon industry. Over the years, the oil and gas industry has learned lessons from a number of process safety events and various hazards such as loss of control and loss of containment top events that could result in: fires, vapour cloud explosions, toxic and asphyxiant exposure, which occurred in various parts of the world and are not limited to the country of origin. The design and operation of process facility for the refinery and petrochemical complex have been the subject of major hazard to the oil industry.

It has been the continuous effort of the industry to update the design and operating philosophies to improve the inherent safety of the processing units. The present paper is concerned with defining the major safety considerations of international oil companies, and Industry Standards in designing and operating the refinery and petrochemical process facilities. Inherently safe design and structured project assurance and audit program are mandatory to guide and drive the design intent of HSE critical systems. The main safety philosophies include: ESD and blowdown, fire and gas detection, toxic gas (H2S) detection, active and passive fire protection (including guidelines for firefighting systems and fire protection), layout HSE philosophy, process safety and containment etc.

This paper provides the **assurance/audit framework** for a major project but does not attempt to cover in detail every aspect of safety design. Reference should be made, therefore, to specific discipline philosophies and specifications as appropriate. (Refer to **Figure 1**).

Project audit or assurance program shall be applied throughout all the project development phases, engineering, construction, start-up, and operation via dedicated and well-structured efforts. The safety philosophy, and resulting protection of persons and assets, is combining all provisions undertaking during all project phases. (Refer to gated process, 1-6 **Steps** explained below).

HSE Design Philosophies

During the detailed engineering, and subsequent construction and commissioning, particular care will be taken by achieving:

- Good engineering practice
- Application of an all-encompassing QA system
- Development and implementation of a safety management system
- Design reviews at 30%, 60% and 90%, layout review, HAZID, Design Safety Review (DSR), accessibility reviews and other reviews.
- Qualitative and quantitative risk assessments and IPF review
- Design of installation for extreme weather conditions
- Execution of an extensive pre-start up review of the plant before introduction of feed (Hydrocarbon, gas) covering both hardware and, control systems and procedures.

- Conformity check with local regulations, project standards and other applicable industry standards and directives.

Gated Process During Project Development and Project Execution Phases

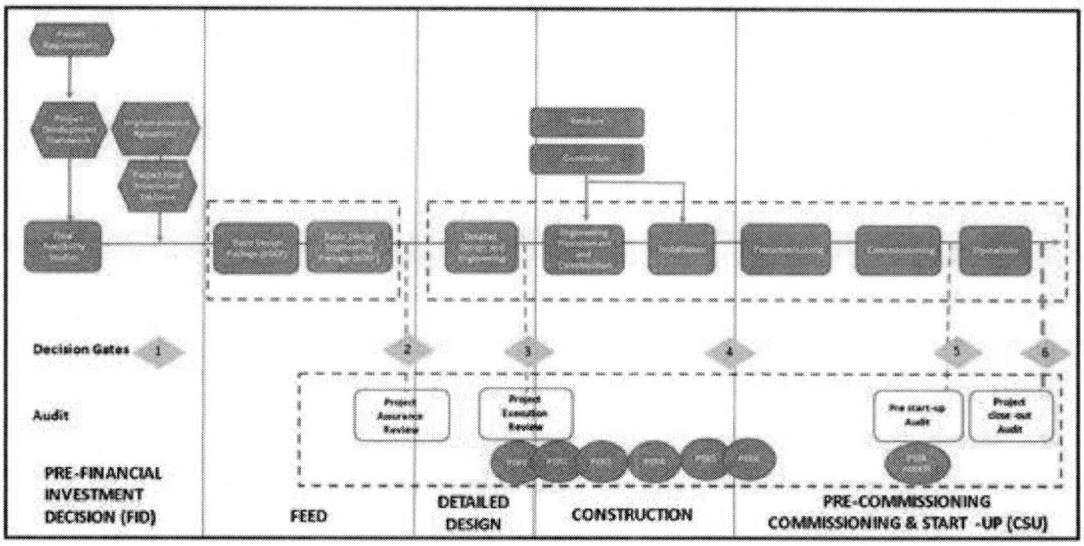

Figure-1 Gated Process for Project Assurance/Audit Framework for Project Execution

Project Assurance Review/Audit is required at all stages of project execution.

1. Pre-Financial Investment Decision (FID) stage:

1) Management: review critical contractor's capability, business license, & project experiences, PEP, schedule.
2) QA/QC: review critical contractor's ISO 9000 certificate, QA/QC plan, standards/policies/procedures (**SPP**), sample check of work records/reports etc. (see Table-1 International Codes and Standard)
3) HSE: review critical contractor's ISO 18000 certificate, project HSE plan, SPP, sample check of work records/reports etc.
4) Technical: review Feasibility Study, Environmental Impact Assessment (EIA) reports., compare critical design area engineering standards against international standards as well as best industry practices, etc.

Deliverable: review comments or audit findings

2. Front-end Engineering Design (FEED) Stage:

1) Management: review critical contractor's capability, business license, & project experiences, PEP, schedule
2) QA/QC: review critical contractor's ISO 9000 certificate, QA/QC plan, SPP, sample check of work records/reports etc.
3) HSE: review critical contractor's ISO 18000 certificate, HSE plan, SPP, sample check of work records/reports, including **PHAs/HAZOP/QRA** etc.
4) Technical: review FEED engineering design, & risk assessment e.g. HAZID, HAZOP, QRA, etc.

Deliverable: audit findings on Statement of Review for FEED, Statement of Review for FEED Safety Study.

3. Detail Design Stage:

1) Management: review critical contractor's capability, business license, & project experiences, PEP, schedule

2) QA/QC: review critical contractor's ISO 9000 certificate, QA/QC plan, procedures, standards/policies/procedures (**SPP**), sample check of work records/reports etc.
3) HSE: review critical contractor's ISO 45001 certificate, HSE plan, standards/policies/procedures (**SPP**), sample check of work records/reports etc.
4) Technical: review Detail Design engineering design, design variation, & risk assessment e.g. closeout of HAZOP recommendations, etc.

Deliverable: review comments or audit report on Statement of Review for Detail Design, Statement of Review for Detail Design Safety Studies.

4. Procurement Stage:
1) Management: review critical contractor's capability, business license, & project experiences.
2) Technical: review specifications of safety critical equipment and QA/QC process.
3) Engineering requirement e.g. specification, database, drawings etc.
   - Manufacturing documents e.g. drawings, inspection & test plan, witness Factory Acceptance Test (FAT), review FAT report & Manufacturing Databook/Spare parts, etc.
   - Inspection & Test Organization documents e.g. Inspection Report, 3rd party certificate etc.

Deliverable: review comments or audit findings.

5. Construction Stage:
1) Management: review critical contractor's capability, business license, & project experiences, PEP, schedule, etc.
2) QA/QC: review critical contractor's ISO 9000 certificate, QA/QC plan, standards/policies/procedures (**SPP**), sample check of work records/reports etc.
3) HSE: review critical contractor's ISO 45001 certificate, HSE management system, plan, procedures, standards/policies/procedures (**SPP**), sample check of work records/reports etc., construction environmental monitoring, waste treatment plan, emergency response plan
4) Technical: compare critical design area project engineering standards against international standards, review construction methodologies, check list, punch list, site construction/safety/environmental/QA/QC activities, design variation, mechanical completion, handover report etc. (**see Mechanical Completion Certificate Checklists** for conducting site audits

Deliverable: review comments or audit report, Statement of Review for Construction

6. Pre-Commissioning & Commissioning Stage:
1) Management: review pre-commissioning & commissioning plan schedule, integration plan etc.
2) Technical: review pre-commissioning & commissioning plan, isolation list, punch list, Site Acceptance Test (SAT), red-line markup drawings, design variation, witness pre-commissioning & commissioning activities, Prestart-up Safety Review (PSSR), ready for startup inspection etc.

Table – 1 International Codes and Standards

IEC 79	Electrical Apparatus for Explosive Gas Atmospheres
API RP 520	Recommended Practice for the Sizing, Selection and Installation of Pressure Relieving Devices (Part 1 and 2)
API RP 521	Recommended Practice for Pressure Relieving and Depressurizing Systems
NFPA 10	Portable Fire Extinguishers
NFPA 11	Low Expansion Foam
NFPA 13	Installation of Sprinkler Systems
NFPA 15	Water spray Fixed systems for Fire Protection
NFPA 16	Deluge Foam Water Sprinkler and Foam Water spray systems
NFPA 17	Dry Chemical Extinguishing systems
NFPA 24	Installation of private fire service mains and their appurtenances
NFPA 30	Flammable and Combustible Liquids Code
NFPA 72	National Fire Alarm Code
NFPA 101	Life Safety Code
NFPA 2001	Clean Agent Extinguishing System
NFPA	Fire Protection Guide to Hazardous Materials
IP Model Code of Safe Practice Part 3	Refinery Safety Code
IP Model Code of Safe Practice Part 19	Fire Protection at Petroleum Refineries and Bulk Storage Installations
BS 5839	BS 5839 Part 1988 for Fire Detection and Alarm Systems for buildings
IEC 61511	Functional Safety, Safety Instrumented Systems for the Process Industry
ISO 14692	Petroleum and Natural Gas Industries – Glass Reinforced Plastics (GRP) piping

This paper discusses industry experience and summarizes the salient design considerations and risk assessment to improve safety integrity in the execution of capital projects for refinery and petrochemical complex. An effort has been made to highlight the ***project assurance/audit framework*** *with HSE methodologies adopted by major oil companies using project standards, and industry best practices and standards like API, ASME, NFPA, IEC, ISO etc. Emphasis has been made to increase the awareness among the discipline design engineers and process safety engineers to develop safer designs of the refinery and petrochemical facilities*

619336 Case Studies – Beware of Common Errors during Risk Assessment of Powder Handling

Abstract:

Case studies – Beware of Common errors during Risk Assessment of Powder handling

"Single size shoe does not fit all," sounds so much in synchrony when we want to assess risk related to powder handling. Powders, unlike liquids have varied challenges right from it's transportation, receipt storage and handling.

Dust handling can have primarily 2 major hazards:

1. Dust explosion
2. Dust thermal decomposition

Few of the important parameters to understand for dust explosion are:

1. Fuel characteristics of the powder:
2. MIE: Minimum energy that can ignite dust cloud in the air.
3. MIT: lowest temp of a hot surface that will cause dust cloud (not dust layer) to ignite.

- Particle size: Smaller size poses higher risk

1. Kst: Max Rate of pressure increase. lower the kst number means the lesser severity 0 bar m/s means no risk while >300 bar m/s means high risk and bigger explosion severity.
2. Ignition source:
3. Hot Surfaces
4. Mechanical sparks/friction

- Electrical motors/contacts

1. Static electricity

Few important parameters to understand and control thermal decomposition in powders is:

1. Onset temperature of decomposition
2. Ageing of powders in vessels that the in process powders are handled such as Dryer, pipelines etc
3. Bulk storage stability of powders

Powders have challenge in terms of dissipation of heat, sampling of representative samples and uniform temperature indication and thus control.

The case studies will focus more on common errors during defining of controls for powder handling and will focus on how preventive controls can be built in to avoid major incidents.

Sr.No	Case Study Title	Controls in brief
1	Drying of Powders with detonation property	Selection of right type of Dryer to ensure metal tramp is not an issue Inert atmosphere
2	Aging of powders and thus lowering of thermal decomposition onset	Thermal studies post commissioning to ascertain periodic cleaning of process vessels such as dryers to avoid decomposition at lower temperatures
3	Handling of shock sensitive powders	Handling the powders in situ in inert atmosphere Ensuring that all process lines such as filter to dryer are flushed every batch Higher authority control for permit to work on such systems Avoid use of agitators for such processes.
4	Handling low MIE powder formulations including drumming out	Static charge generation is a concern. Some preventive controls to be considered are: Ensuring that drop height of powder is less than 3 mtr. Use of earthing continuity stem such as earthrite system. Ensure personnel use antistatic body protection and shoes. Use of conductive FIBC or drums

Good housekeeping

All the above studies are based on Author's 10 years of experience in process hazard analysis for bulk chemical manufacturing and is expected to benefit the audience at large.

BS Team: Finding Competence Gaps

Salvador Ávila
UFBA
avilasalva@gmail.com

Luiz Fernando Lopes de Souza
COPENOR
lflsouza@globo.com

Gunther Josua Costa
UNIVASF
guntherjosua@gmail.com

Lucas Menezes Pereira
UFBA
lucasmpereira13@hotmail.com

Prepared for Presentation at
American Institute of Chemical Engineers
2021 Spring Meeting and 17th Global Congress on Process Safety Virtual
April 18 - 22, 2021

BS Team: Finding Competence Gaps

Salvador Ávila
UFBA
avilasalva@gmail.com

Luiz Fernando Lopes de Souza
COPENOR
lflsouza@globo.com

Gunther Josua Costa
UNIVASF
guntherjosua@gmail.com

Lucas Menezes Pereira
UFBA
lucasmpereira13@hotmail.com

Keywords: BS, Competences gaps, Decision models

Abstract

The academy form competences to work in the routine, without worries, without unknown hazards. The rules and objectives in routine activities are clear and require operational discipline to control the cost of production and product quality. The current educational model does not prepare society and groups for an emergency or even for decisions in a crisis, where resources are scarce mainly because the necessary knowledge about danger and its mechanism of action is lacking.
What really happens in the chain event scenario when the danger is unknown? What are the typical models of action of the emergency leader and his team? What are the required actions in crisis management required for a BS type event considering the scarce resources, the high level of stress and the direction of the danger energy? Which models are indicated for the situation of unknown danger and its intense energy flow with high impacts during these events? What is the appropriate mind map of those who lead and execute the actions considering the low level of visibility in the events and the dynamic of geopolitics in the BS scenario where, depending on the preparation for emergency, in case of intense demand for attention and action, possible modifications may occur in the map mental for decisions?
This work intends to review the discussions already performed about new concepts and tools to be used in the crisis prevention area. Between this works we will understand the effect of the stress level on emergency decisions valuating human error and observing team response using LODA tool. The Human elements designed in the organizational culture, can stop the high danger energy that can flow from some industries. The principal aspects to be constructed and monitored are technology (risk and complexity); management (leadership and stress); and behavior (cooperation, commitment, competence, and communication).

By the other hand, human factors are analyzed to avoid situations which hazard energy carriers during top events, as the effect of cultural aspects to chain reaction until the occurrence of disaster. Decision models for the emergency brigade (leader and team) indicate the motor and cognitive gaps that cause lack of control during contingencies. Finally, the analysis of simulation types, by rotating observation of field simulation allow find the deficiencies of the emergency team in a real situation in the future.
The methodology divided in following steps are dynamic: first, rescue the discussions about stress and the BS events and, in second step, recognize the intense hazard energy flow through failure events, antecedents, elements and human factors. In third step, we try to locate the gaps and environmental conflicts influencing losses from accidents. Finally, we assess the current event of covid-19 pandemics to suggest correct emergency preparedness actions in the preparation of leaders and their teams regarding cognitive and motor characteristics in situations of high demand. When comparing the characteristics already described by TALEB about BS with the event COVID-19, we intend to carry out a case study to indicate the possible causes this global impact crisis, or even, understand the technological disaster causes where uncontrolled events from chemical, nuclear, and oil industry begins chain reactions until the disaster. Unfortunately, a Crisis situation can be caused from several intricated dimensions, making difficult the contingency and mitigation actions.

1 Introduction

The academy form competences to work in the routine, without worries, without unknown hazards. The rules and objectives in routine activities are clear and require operational discipline to control the cost of production and product quality. The current educational model does not prepare society and groups for an emergency or even for decisions in a crisis, where resources are scarce mainly because the necessary knowledge about the hazard and its mechanism of action is lacking.
The knowledge learned to graduate the engineer is limited to logic and objective concepts. The students do not discuss different concepts, as social culture, legislation, workstation behavior at the same time, mixing objective aspects with subjective ones. We need to understand Cultural traits, geopolitical aspects, intrinsic variables from worker during routine to discuss about response models for emergencies, necessary to avoid disasters.

Unfortunately, we only look for new competences after the disaster happen! Why?

The metrics for Process Safety management are discussed in methods indicated in the CCPS books [1]. Some important issues such as: management and investigation of major incidents [2]; technological changes that cause accidents [3]; lack preventive actions for organizational changes [4] are exemplified and barriers are suggested, so as not to affect the safety of processes with big accidents.
These issues and solutions seek to avoid the accident through management actions, technological systems, and the attempt to control human factors [5] based on the lessons learned by various industries and associations when investigating the events that occurred. This investigation of the past, looking back to the facts, has a limit when external, dynamic environments threaten normality. The crisis scenarios can be very dynamic, not visible and of parallel occurrence, at the same time as those caused by climate change, by social conflicts, by conflicts arising from

degraded geopolitics, and by the effects of multiculturalism in the world of work. and in society - indicating low tolerance behavior.
In 2018, CCPS [6] with the leadership of Eastman, raises the issue of normalization of deviation, indicating repeated situations of acceptance of operational deviations, in the shift routine, are considered normal situations. The lack of research with the indication of bottom-up techniques for the analysis of operational culture and the respective heuristics turns the book into another attempt to guide the occurrence of events looking to the past and without an appropriate view of what is happening in the present. and what will happen in the future, on the factory floor. What are the social emergent that cause deviations to normalize?

What are the social emergent that cause deviations to normalize?

In this context, analyzing the internal and external social threats to avoid environments prone to deviations gains importance. Risk aversion, excessive self-confidence, and fear of guilt for human error are considered cultural events that produce underreporting. This phenomenon changes the shape of Bird's pyramid [7] to the Christmas tree as shown in Figure 1 and causes the accident to occur already with the chain reaction triggered. The assessment and actions resulting from the notification (not performed) could avoid the domino effect or reduce the risk of major events.
The accumulation of deviations without treatment and with high level of underreporting, a cultural event, increases the likelihood of Black Swan events. Failure to recognize the Scenario for hazard identification and risk analysis that implies a lack of insight into the operational context and creates uncontrolled situation with natural, technological, political disasters. The Black Swan Event goes beyond the local scale and reaches a global scale, with unknown dangers, or already known dangers, but due to lack of attention, it has the characteristics of unknown danger because it has not been investigated as to the mechanisms of action. We do not know what the appropriate barriers are. Thus, we classify the Black Swan because of totally unknown or operationally unknown dangers.

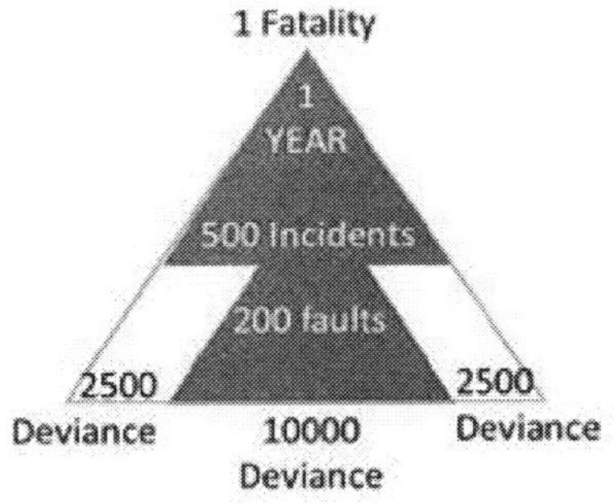

Figure 1. Christmas Accident Tree

2 Literature Review

This work intends to review the discussions already performed about new concepts and tools in the crisis area. Between this works will understand the effect of the stress level on emergency decisions valuating human error and observing team response using LODA tool. The Human elements in the organizational design can stop the high hazard energy flow of some industries. The principal aspects to be constructed and monitored are technology (risk and complexity);

management (leadership and stress); and behavior (cooperation, commitment, competence, and communication).
By the other hand, human factors are analyzed to avoid situations in which hazard energy carriers during events, from cultural aspects to chain reaction until disaster. Decision models for the emergency brigade, leader, and team; indicate the motor and cognitive gaps that can result in lack of control during contingencies. Finally, the analysis of simulation types, by rotating observation of field simulation allow find the deficiencies of the emergency team.
Given the great stress present in the geopolitical situation, in which there is a constant struggle for territory, renewable and non-renewable energy sources, in addition to climate changes intensified by anthropic activity, large flows of Hazard Energy (HE) are generated that are commonly not perceived. The unnoticed concentration of this energy gives rise to what is known as latent hazard, and when it evolves to levels that can no longer control or filter its pending of burst and destruction, a megaevent of disaster occurs, which has recently been called the Black Swan.
Due to the lack of a notion of the hazard to which disaster victims are subjected, it is said that the Black Swan is still unknown to the population, and because they are often unprecedented occurrences, they are considered rare and of high impact in multiple respects. For [8], the victims of the twin tower attack on 9/11 were victims of a Black Swan. It was not possible to locate where the hazard energy was concentrated, nor measure it. Gray Swans, according to the same author, are unlikely but impactful events that are outside the normal scenario.
A Black Elephant, according to [9], is a significant threat, risk or issue that is obvious to everyone, but no one wants to address or deal with it, "they get ignored". When it causes high impact, people get surprised and act as if it were an unpredictable black swan.
There are new models of behavior to include high stress and ways to measure this feature through the radar chart. This chart presents the characteristic of presenting multivariable data in the form of a graph.
The managerial, human-social, and technological aspects can create an environment for unsafe behavior, in which they allow the flow of hazard energy across barriers. The human and organizational factors elaborate, by events, the condition for the transfer of the social hazard towards the physical realization of the accident or incident.
In order to prepare people to deal with risk situations and be able to participate in decision-making processes that report to their own lives, it is necessary investment in: 1) Integrated Communication - Training and language competence are really important. Subjective factors such as judgments, culture of guilt and fear of damage to the image cause the worker to omit information that could contribute to the construction of barriers to avoid incidents and accidents; 2) Stakeholders - person or group who have an interest in a particular business, industry or service; 3) Intellectual Capital - Knowledge of preventive and mitigating actions provides people with adequate decision-making; 4) Competence - Understanding the causal nexus of the phenomena involved in the accident, having experience and experience similar to the activities related to the events.

3 Models & Hypotheses

The discussion about the competence of Black Swan Team beginning answering questions after observing real situations of emergency. After we feel some gaps and try to answer after investigation about the reasons of emergency team perform this behavior. After each question we construct or explain the scenario and answer with new tools that will compound the methodology.

3.1. What really happens in the chain event scenario when the danger is unknown?

We have nothing to perform about the existence of a disaster from an unknown danger as in actual case of a pandemic. The disaster is a truth, we must reduce its impact, investigate the various dimensions involved with this unknown danger, or unknown operational danger considering the cultural and environmental effects during the isolation and flexibilization.

The study of the origin of the disaster allows us to understand its mechanism of action and possible consequences. The danger that causes the accident-disaster of technological (industry) or natural (climate, virus) origin can be acute, being a fast event with high impact, or it can be chronic, lasting a long time with, probably, immense impacts. Thus, the event becomes a public accident when it affects the population. We classify the danger and consequent events of COVID19 as natural, chronic, and public. This discussion is part of Strategic Intelligence that define the need resources and Hazard Identification, and Analysis of General Risk to define and test the best barriers.

While the danger is at work in an external region or country (not close to our region) we will observe and learn about its mechanisms, we seek to install two instances: The Integrated Management Committee and the Strategic Intelligence Committee. The first will join the necessary resources and leaders to take decision and the second will look for the best actions, benchmark, learning from other countries and regions.

In Integrated Management, Specialists, Managers and Practitioners who are part of Public-Private Institutions meet to define studies that assist in the decision on actions. Some tests are suggested to find the mechanisms of action of the hazard. In this way, barriers are built for known dangers, for the unknown and for threatening environments. Indicators and parameters and variables are mined to understand the impact of the pandemic on the economy and social relations.

Strategic Intelligence intends to identify the source of the main hazard and parallel hazards to project the Future Crisis Scenario. Scenarios are designed to test barriers, variables, resources, and knowledge. We will value to learn from the past, and from abroad experience, sending representatives to recognize the mechanism of these dangers through visits to government and institutions. In this moment a good general knowledge about pandemic situation, technological disaster and natural, impacts on social conflicts and economy, will help do draw an initial description and after, inviting specialists, the crisis leader tries to construct and perform effective actions (corrective, preventive and mitigation actions).

After confirmation that the Scenarios indicate an active Pandemic, the Contingency Committees begin in specific and critical regions. In some cases, such as natural disasters involving an earthquake, breaking of barriers in the mining industry and other technological accidents (fire or toxic spill) we need to install a Contingency Committees. In cases of acute accidents, groups are sent for mitigation action. Therefore, operations and local intervention teams will be installed.

The Black Swan Team at the Strategic (Strategic Intelligence), Tactical (Integrated Management) and Operational (Command and operations) levels need the classification of the type of event and impact for the work of the teams. One of the cases may be the study of technological accidents such as the breaking of barriers in mining or the leakage of toxic gases near the city (similar to Bhopal). The other case can be related to public events of high capillarity about impact, a virus that contaminates quickly and through unknown routes, the current COVID19, or even a series of earthquakes that promote tidal waves and affect the functioning of cities. Or even, the possibility of multiple events occurring involving the technological and the public dimension.

3.2. What are the typical models of action of the emergency leader and crisis management required for treat a black swan type event considering the resources, the level of stress and the direction of the danger energy?

LODA (PROGRESSIVE STRESS MULTI-LYER MODEL) shows that the mind map change in extremely high stress with wrong decisions.

Ávila [10], knowing about the possibility of high impact of technological accidents near cities and after confirming the risk to the risk affected by the work team and the population through underreporting. Ávila confirms the occurrence of the phenomenon of cultural normalization of deviation and for the analysis of the consequence of progressive stress, he designed an accident involving an LPG fireball near a city in Brazil.

This event constructed from real events and deviations that occurred, although the top event did not occur, was composed of 65 diversions and successive events, sociotechnical failures in the form of the local routine and complemented with those that occurred in New Mexico. The Accident prepared for the Risk Analysis impacted with 40 fatalities (this simulation was structured in the thesis and published as a tool [10].

The analysis model involved 5 levels of stress in the team, where 1st is uncontrolled process with operational deviations and the last was the crisis scenario in the next city. Progressive stress disperses or disorganizes the mind maps that affect the emergency leader's decisions. In this model it is considered the occurrence of a theater with critical roles where there are communication failures or non-acceptance of the roles in the staging. Observing each phase of the accident and its cognitive stress will indicate social phenomena that represent problems in the treatment of accidents. These problems must be addressed before the actual demand for contingency.

EMERGENCY PERFORMANCE MODEL indicated that there are characteristics for the leader and the team that are intuitive, affective and cognitive and that it was not found in emergency team in these accidents of the history.

The studies by Carvalho [11] and Ávila et al [12] indicate that there are specific characteristics that are appropriate for the leader and the emergency team on real demand. In the case of leadership, we can mention perception of real risks (signs and scenarios); competence regarding the danger mechanisms (fire, virus, mechanical impact); energizing the team to act in contingency; a good memory to reinforce the decision; and speed of command since the time of the chain event, until the greatest impact, is very fast, and evacuation measures can save many lives.

In the case of the emergency team, we can mention: the quick cognitive to retrieve documents, a discussion, a phone call and assist the leadership in the decision through opinion and calculations; quick rescue of memory for the analysis of the scenarios, this aid sometimes brings information avoiding the direction of the danger to where there is greater vulnerability, for example, sending of chemical product to empty the burnt tank where there are visitors or maintenance service - the leader does not knows the level of vulnerability of the areas and depends on the team's support for this analysis; emotional balance brings a better quality of the cognitive and motor apparatus; ability to predict future situations through greater knowledge of the event's phenomena; and the ability to act on time, without delay, bringing firmness to field interventions.

Thus, Carvalho [11] and Ávila et al [13] affirm that the preparation of the emergency team mixes past experiences, attitudes, field and memory exercises, team energization, which are often not considered in normal training for teams of emergency. The results of the team's performance must be observed so that there is continuous improvement in performance under real demand.

COGNITIVE GAP ASSESSMENT IN REAL ACCIDENTS AND SIMULATION where we assess the emergency teams based on public and technologic accidents confirming deficiencies indicated by Carvalho, Ávila e Pichon Riviere.

This research work analyzed the national and international accidents regarding the characteristics discussed in the models for emergency action [11] [12]. The emergency team was analyzed in Accidents: (a) Chemicals and Oil of high impact in number of fatalities and environment (piper alpha - oil; Horizon - Oil; Union Carbide Bhopal - isocyanate); (b) national accidents of shipwreck in Mar Grande and explosion with fatalities at the FPSO São Mateus in the Brazilian state of Espirito Santo; and accident designed in the form of a simulation for the analysis of fireball risk by LPG in northeastern Brazil.

Specialists working in Risk Analysis, Contingency and Emergency Leadership met to classify the teams in terms of cognitive gaps and emotional control. This analysis was published at LACPS in Buenos Aires and it was concluded, in discussions with Process Safety Managers, that the classic way of training for emergency brigade teams does not meet the need to control real demand. Why? Although the situation presented in the training simulations for the Emergency Brigade, the petrochemical, refining, and oil industries seek typical situations for fire, fire, leaks and environmental impact, there are always organizational factors that affect human factors, there are always social factors that they cause conflicts and affect the quality of communication at work, and there are always false indications, a high backlog in the industry today. Thus, events always happen in multiple dimensions and the possibility of something unknown happening is real. Thus, the greatest value to be given is cognitive-affective-intuitive training for the emergency brigade. Lately, Process Safety Managers work only in Cognitive and without surprises. The level of stress and the style of the Leadership bring a level of comfort that may not be satisfactory for the occurrences of real demands.

COGNITIVE FAILURE IN DISASTER FIELD SIMULATION where the team is not always integrated and energized in the hot simulation situation, believing only in the obligation to be present. In this case, preparation through the simulation leaves cognitive and compromising gaps for the contingency in the disaster.

At the CIDEM, Mass Disaster Congress, held in Feira de Santana, UFBA was a rotating observer and investigator during a simulated exercise involving 25 public agencies including Armed Forces (Navy, Army and Air Force), Federal Police, Municipal Guard and State Military Police, SAMU, State Fire Department in addition to the BOPE anti-terrorist squad. Various functions of civil defense, social defense, isolation, and medical care came together in this exercise to prevent events of bomb explosion on buses, tipping of benzene trucks, and kidnapping, actions promoted by terrorist groups, all with an impact on the population and in nature, in the road modal.

The observation of these events indicates that the hierarchy of the agencies is above the recognition of their roles during the simulation and the inter-agency communication. The communication made with noise leads the team to different levels of urgency for mind mapping in the different teams. Some equipment was underused indicating lack of training, therefore, there are human and organizational factors that must be put into discussion to improve the integrated actions of the emergency command. The main characteristic is the lack of common priorities in the agencies promoting a disintegrated management state, where the size of the problem is not known for sure, only when faced with it. We note mainly that communication tools and devices are not used properly. Because it is a cold simulation, prescribed events of common knowledge, many communication failures did not appear. And this is a serious problem when analyzing previous events when municipal, federal, state forces and private governance act around disaster mitigation, for example, by preventing ship collisions. We note that the lack of credit of the great emergency teams on the simulated exercise can induce people to repeat mechanical actions bringing nullity to the Strategic Intelligence team.

3.3. Which models are indicated for the situation of unknown danger and its intense energy flow with high impacts during these events?

SYSTEMIC FAILURE OR FINGERTIP FAILURE MODEL, where the failure is multidimensional and involves from emotional balance, and knowledge - quick cognitive, we must analyze the connection between these properties to invest in selection and training.

The systemic failure models that indicate connectivity points of events of different dimensions to cause the top event. The format of the Systemic failure is like the digital of the fault that was published at the European Reliability Congress (ESREL).

EXERGY OF OPERATIONAL FAILURE AND COMPLEX PROCESSES where the indicators of complexity in the construction of BS are analyzed.

The discussion on the Exercise of Operational Failure [14] presenting that cognitive losses in the task are related to energy losses in equipment and processes. The presentation of The Investigation on Complex Processes through the tools that connect different sources that cause uncontrol in the process, interpreted through: Connectivity Diagram [15] [16]; of the Materialization Diagram where the failures that transform communication noises from management into a disconnected mental map in execution, are transferred into disturbance in the physical dimension; and finally the logical diagram of the fault indicating regions where to better install the barriers, in sociotechnical functions. These are models discussed at the Brazilian Congress of Chemical Engineering and applied in the Chemical Process Industry (aliphatic and aromatic amines in addition to liquefied petroleum gas).

NETWORK OF FACTORS AND HUMAN ELEMENTS, BARRIER EFFICIENCY, AND DANGER ENERGY SATURATION where we need to identify the points where danger energy enters saturation to elaborate the best barriers.

In recent times we have discussed ways to identify the regions in which the greatest flow of danger energy occurs, following the Model of Reason's Swiss cheese [17]. This danger flow circulates among the factors that influence human performance on the sequential, parallel and crossflow route. This multidimensionality of the failure follows the model in a network format of human and organizational factors divided into culture, management, routine and accident. This multidimensionality of the failure also discusses the safeguards of the technological, managerial, and behavioral dimension through the human elements that are design barriers to contain the danger energy that flows until reaching the maximum levels that are the accident and the disaster. Human factors are part of the routine of the operation and human elements are design safeguards. As described in Table 1, in the operation there are different factors that cause incidents and situations of loss. The nature of this event has the following dimensions: Cultural; Managerial; Routine control; and the scale of the accident and disaster. Also, in the same table we present the types of human elements in the project: Management; Technology; and Behavior.

Table 1 – Human Factors and Elements

OPERATION - A. CULTURAL DIMENSION OF HUMAN FACTOR: (1) Social Culture - (2) Social Phenomenon - (3) Organizational and Safety Culture. B. MANAGEMENT: (4) Management - (5) People & Groups. C. DIMENSION OF CONTROL IN THE ROUTINE: (6) Operational Control - (7) Failure Control D. ACCIDENT AND DISASTER DIMENSION: (8) Accident Control - (9) Disaster Control.	**DESIGN** - E. HUMAN MANAGEMENT ELEMENT (1) Stress - (2) Leadership F. TECHNOLOGY (3) Risk - (4) Complexity G. BEHAVIOR (5) Communication - (6) Competence (7) Cooperation - (8) Commitment

4 Methodology

The investigation was carried out following the steps: (1) Recovery from the discussion about stress in the treatment of emergencies, especially in the case of BS type disasters; (2) Recognition of the intense flow of danger energy through antecedent events, human safeguard elements, human factors of industrial operation and consequent events that hinder the return to normality; (3) to locate the environmental gaps and conflicts that influence process losses and the occurrence of accidents, chain reactions and the difficulties of returning to normality, for this we divided the competences in: Strategic Intelligence, Integrated Management, Command and action in Local Operations, and Organizational Resilience. For each competence, possible gaps, deviations, or human errors are discussed, comparing with the expected standard.
Finally, we assess the current event of covid-19 pandemics to suggest correct emergency preparedness actions in the preparation of leaders and their teams regarding cognitive and motor characteristics in situations of high demand.
Here in the methodology, we will indicate the competencies required for the BS team and the mental map of each competency for further discussion on the case of the COVID19 disaster.

4.1. Competencies for BS

As shown in Figure 2, Strategic Intelligence and Organizational Resilience have the purpose, respectively, to identify the danger and to restore the organization to the condition of normal, controlled operation, as it was before the disaster. Therefore, they outline strategies based on investigations of the past, the present, and the future, in addition to comparing local situations with external ones and in other countries. It is important to identify what the danger is and its mechanism of action, map the impacts to project the recovery. The Intelligence group is composed of risk analysts who work with scenarios of different dimensions (health, politics, technology, climate) and who can defragment dispersed information to build Future Scenarios and explain the events, even if the danger is still unknown, through testing and observation.
Another group that will guarantee the application of resources in the right place and that will effectively communicate the current situation involving the population and cities, industry, commerce, and services, is the integrated management group that will develop appropriate indicators for decision making from the disaster impact curve and from the data and information generated by strategic intelligence.
The local operations and command group works to carry out the activities indicated by the integrated and recommended management group, guided by strategic intelligence. The role of the

leader is important to energize the team and the role of the team providing cognitive and motor support in this achievement.
In this way, the Academy, Public Agencies and Risk activities such as the chemical industry and the oil industry need to structure themselves to prepare competencies that know how to act in the investigation of scenarios, in integrated public-private crisis management, in command and in carrying out operations. contingency and the recovery of local and regional activities.
Cold simulations [18] do not bring the expected result to structure contingency teams. Crisis Managers have not been effective in the application of resources, in communication and in the strategies drawn from the decision tools, as presented in the pandemic of COVID19 [19]. The risk management team is not able to design the scenario because it does not identify the danger and its mechanisms and impacts. We have no skills prepared for Organizational Resilience and we lose control of the situation during the emergency period. Thus, we need to develop the skills mentioned in Figure 2 and confirmed from the team's mind maps.
Mind maps are average informal rules accepted by the group after all the list of organizational priorities, group preferences, and regionalities. This map will then show how the team works in each of the four blocks of competence and how the teams should work.
Note that the models do not characterize the relationship of urgency of time, recognition of danger and its mechanisms, also the importance of the discussion on the types of communication (between groups, with the community, with the population or inserted in a political environment) and care in relation to false-incomplete or anticipated information.
The final analysis is a brief discussion of the mind map for BS events in general and for the specific case of the COVID 19 Pandemic.

5 Case Study

The Research Group on Operational and Dynamic Risks (GRODIN) has prepared a Master Program for Disaster Preparedness and Crisis Management and has been acting as a critical observer of Government actions and people's reactions to address COVID19, discussing tools in the Strategic, Managerial and of Operations.
Unfortunately, new strains of this virus are emerging, in a transformation of unknown origin, which shows once again that we are during a Black Swan Event. We must deal with this matter coldly to act in the face of the terror of mass death, and to avoid the wrong decisions. The skills presented in Figure 2 must be developed and available to act in this Global Crisis:

(1) IE Strategic Intelligence (study of the origin of the danger, its mechanisms, and the construction of future scenarios, where spurious data and warnings about changing the danger line must be valued, investigating the past to understand the present, investigating the future for disaster preparedness.
(2) Integrated Management and Crisis Project - GIP (data flow, organization of resources, development of tools for decision appropriate to the mechanisms of action of the danger and its dynamism when the danger is unknown, communication mode and the treatment of the sub- political-economic and technological notification).
(3) Command of local operations COL with the preparation of resources for the here and now based on the organization of the Integrated Crisis Management in the projections of the future created by the IE (people, structures), reviewing the Integrated Management and the IE with new data from the present coming from the COL, having a preparation of the leadership and the team to avoid simulating extreme stress.
(4) Organizational Resilience RO where, with the control of the situation, seek the causal nexus based on practice and science in order not to lose control of the recovery. Always value spurious data

and information and warnings, classifying the quality of the information. Motivate IE for further investigations about the danger, its mechanisms of action and possible impacts before it happens.

Taleb [8] stated that in the face of an unknown danger, "everyone" will try to explain without knowing for certain what he is talking about (we need IE), demonstrating dominance that does not exist. Others who are comfortable with being a subject of scientific knowledge, virology, are not concerned with the preparation for situations not lived in the past, so we consider that unknown danger = known local non-operational danger. Thus, with chronic and growing impacts, we will have uncontrolled scenarios due to lack of resources, infrastructure, and lack of competent people for action.

The Black Swan has several dimensions that occur at the same time, Health, Economy, Politics, Logistics, Climate and Geopolitics, regardless of the dimension of origin of the danger. This multidimensionality gets in the way of decisions since the leaders of knowledge, public and private management, do not have the habit of thinking, investigating and deciding considering the systemic failure and its multidimensionality.

The low quality of information, underreporting and / or addictive data, undermines the role of management and science, which creates addictive or incorrect interpretations following mathematical models that are not practical. Observation of facts and events must be valued in times of Crisis (Disaster) to enable the endless elaboration of multidimensional scenarios to dynamize the Strategies.

In times of crisis, it is necessary to work with two teams that feed and feed each other on information. The IE and RO must be calm and when analyzing fatality data, review future actions and investment of restricted resources. The GIP and COL that see in the here and now practical needs to control the mitigation teams towards operational control and human errors caused by extreme stress (EE).

The IE and RO must be calm in the construction of Scenarios with weekly average of 4-day moving averages of fatalities to assess the effectiveness and effectiveness of previous actions, seek to explain or approach the truth about the danger. IE and RO's lack of calm in analyzing the scenario can vitiate interpretations in a single dimension of the disaster leading to incorrect projections of the future. It is necessary to change the position of the gaze to see what is hidden. Thus, we have developed guidelines to be followed:

(1) Installation of Crisis Committees for IE, GIC, COL and RO with a multidisciplinary and dynamic team; (2) Choosing-developing easy-to-use and effective statistical tools to support decisions, therefore, the weekly average of the 4-day moving average tries to avoid underreporting and spurious data that does not indicate a trend; (3) Enhance research on spurious data or information or warnings that may indicate a change in the danger or its mechanism and, consequently, its impact; (4) Prepare the team to maintain the quality of the cognitive and also adapt to the "time available" to guarantee the effectiveness of the actions. Extreme stress increases human error from 10% to 70%, bringing incorrect decisions and actions. We have to use tools to preserve the cognitive in extreme stress !; (5) Build a knowledge network platform, develop practical tools based on grounded theory, information flow, integration, and critical observation; (6) Following the film World War Z, it is important to treat spurious data and observe-validate unconventional local solutions that can be the difference in the application of restricted resources; (7) Following the UTOPIA Series, recognizing the political, geopolitical movements that have low visibility, can indicate simulacra of information, fake news. When knowing the fake news and underreporting, the safeguards will be more assertive, the aim is to test the hypotheses to protect the population; (8) Even if the situation is more optimistic, maintain recourse to pessimistic

scenarios, considering the opinion of people who are reliable but who project a different future from the majority as in the film World War Z.
I forgot the following aspect about the Leader to avoid Accident-Disaster Situations discussed by Collins [20]: (1) Not wanting the power and nor having vanities of signatures; (2) Extreme humility in dealing with unknown hazards; (3) Active listening for events, warnings and deviations; (4) Fierce determination in decisions and actions; (5) Recognize the possibility of own and team mistakes in strategy and action; (6) valuing the organizational and family unit.

Figure 2. Competencies for the BS Team: Intelligence, Management and Operation.

5.1. Mind Map for BS and COVID19

The characteristics already described by TALEB and Avila using the models from research about the dynamic risk we show the principal competences to attend a BS event. In Table 2, we discuss about the generic pattern and a case of COVID19 BS team.
The informal rules that turn into actions follow mind maps by type of function, Intelligence, Management and Operation. Below are examples of expected patterns and deviations observed in different mind maps.

Table 2. Mind Map for BS

Mind Map	Generic standard	Expected standard (COVID19)	Deviations (GENERIC)
Intelligence	- To study the accidents and incidents;	- Study the disease in depth knowing the form of contagion, consequences, forms of treatment;	- Study superficially without exploring for learning;
	- In the crisis, seek as much information about the products involved, the team exposed, the area affected;	- In the crisis, seek as much information as possible about the disease, exposed teams, affected areas;	- Search for a solution without knowing what we are facing, people involved;
	- Know the risks that are relevant to the business;	- Know the risks that are relevant to the disease;	- Not knowing the extent of the risks and not preparing for it;
M an	- Know the resources available to face the crisis;	- Know in depth all the resources available for coping with the disease;	- Have resources that are not available.

			- Wait to seek resources in crise;
	- Treat accidents and incidents with preventive and corrective improvement actions;	- Predict and seek resources to strengthen the fight against the disease with the creation of scenarios;	- Treat accidents and incidents superficially without seeking a solution;
		- Treat deviations with preventive and corrective improvement actions;	- Treat deviations superficially without seeking a solution;
Operations	- Know and study their role in acting during the crisis;	- Know and study their role in acting during the pandemic crisis;	- Facing the crisis in a way that is not consistent with the procedures causing conflict with the team;
	- Communicate any changes that are made to the procedures by training the others involved;	- Communicate any changes that are made to the procedures by training the others involved;	- Carrying out activities in a manner inconsistent with the procedures, putting the activity at risk;

6 Conclusion

The current educational model prepares society for routine operations only, always with a focus on the main objective of producing with quality and low cost. The safety of the process is studied in a superficial way, focusing on the clearest and expected risks. BS are situations that arise during an event resulting from the sum of risks that may have been studied separately but never together. This moment requires several skills from the teams that work on the front line and coping and mastering the situation can take more time and we know that in an emergency, time and knowledge are fundamental. The intention of this work was to present a review and discussion about new concepts and tools that can be used in the prevention of human errors during crises (LODA) and thus we were able to understand the effects that the level of stress can influence in the decision making during the crisis and compromise coping with the crisis based on the understanding that the mental map changes completely in the crisis.

The presentation of the methodology with the necessary skills to face the BS and the mental map models are fundamental to create the PDCA of the crisis that always starts with strategic intelligence, integrated management, Operations command, and the organizational resilience that will feed the strategic intelligence in a learning cycle and the search for continuous improvement. The more we study this methodology, we see that although the event is unexpected, the knowledge and the feedback of information for the knowledge of the risks, will make that the mastery of the situation is reached more quickly.

The academy needs to better prepare society for unexpected situations. BS events are becoming more common, given the COVID19 pandemic experienced today and methodologies applied at the time of the crisis can save lives.

7 References

[1] T. Overton, M. Broadribb, C.N. Cheung, E. Christie, S. Cowher et al. "Guidelines for Process Safety Metrics", pp. 1-178, 2009.

[2] Center for Chemical Process Safety (CCPS). "Guidelines for Investigating Process Safety Incidents", Third Edition, 2019.

[3] Center for Chemical Process Safety (CCPS). "Guidelines for the Management of Change for Process Safety", 2010.

[4] Center for Chemical Process Safety (CCPS). "Guidelines for Managing Process Safety Risks During Organizational Change", 2013.

[5] Center for Chemical Process Safety (CCPS). "Human Factors Methods for Improving Performance in the Process Industries", December 2006, 248 p.

[6] Center for Chemical Process Safety (CCPS). "Recognizing and Responding to Normalization of Deviance", September 2018, 176 p.

[7] Herbert W. Heinrich, Industrial Accident Prevention: A Scientific Approach, 1931

[8] N. Taleb, "The Black Swan: Second Edition: The Impact of the Highly Improbable". Random House Trade Paperbacks, 2010, 469 p.

[9] P. Ho, "Future Beasts and Where to Find Them". Available at https://www.psd.gov.sg/challenge/ideas/feature/future-beasts-and-where-to-find-them. Accessed on January 22, 2021.

[10] S.F. Ávila. "Dependent Layer of Operation Decision Analyzes (LODA) to calculate human factor, a simulated case with PLG event", 2011.

[11] A.C.F. Carvalho. "Diagnóstico de lacunas nas competências da equipe para enfrentamento de situações fora da rotina". Master's dissertation – Federal University of Bahia, Polytechnic School of Engineering. Salvador, 2016, 103 p.

[12] S.F. Ávila., B. Mrugalska, M. Wyrwicka, M. Souza, J. Ávila, et al. "Cognitive and Organizational Criteria for Workstation Design", 2019.

[13] S.F. Ávila., A.C.F. Carvalho, G.P.J. Portela, C. Costa. "Learning Environment to Take Operational Decision in Emergency Situation". Procedia Manufacturing. vol. 3, pp. 1780-1787, 2015.

[14] S. F. Ávila. "A exergia da falha operacional, uma discussão filosófica com bases práticas". In: Proceedings of 19th Brazilian Congress of Chemical Engineering, 2012, Buzios.

[15] S.F. Ávila. "Análise de falha em processos complexos". In: Proceedings of 19th Brazilian Congress of Chemical Engineering, 2012, Buzios.

[16] C. Perrow. "Normal accidents: living with high-risk technologies". Princeton University Press, Princeton, N.J., 1999.

[17] J. T. Reason. "Human error". Cambridge University Press, 1990.

[18] S. F. Ávila., I. Cerqueira, V. Júnior, J. Nascimento, C. Peres, C. Brunoro, P. Arezes. . Integrated Management in Disaster: Discussion of Competences in a Real Simulation. AHFE 2019.

[19] S. F. Ávila., J.S. Ávila, A. A. Santos, M. C. N. Carmo, L. F. L. Souza. Crisis Decision Tool, fighting against COVID-19. AHFE 2021.

[20] J. Collins. Level 5 Leadership: The Triumph of Humility and Fierce Resolve. Harvard Business. 2005.

619165 Case Study in M+a – Communication and Understanding Hazards and Safety Concepts

Abstract:

A multinational corporation acquired a local consumer products manufacturing facility. Hazard awareness and process safety protocols were lacking at the local company, especially as measured against the parent company's corporate standards. The primary root cause was lack of a strong management of change procedures, as the processes in the facility were originally designed for non-flammable products but over time the facility transitioned to handling flammable liquids and combustible dusts.

Hundreds of unique chemicals/products stored and handled at the facility were analyzed to determine compliance with local fire codes and serve as a design basis for safety program improvements. Specific challenges around flammable liquids and combustible dusts were deemed to be highest risk and were addressed immediately.

This case study presents high level findings and both capital and operational budget related recommendations applicable to companies in similar situations. By applying a combination fire/building code evaluation and hazardous materials / process safety gap analysis, the company was able to prioritize risk reduction actions in alignment with their strategic objectives. This work presents a basis for hazardous material safety in facilities that store and handle a wide range of hazardous materials and aligning risk remediation activities with corporate objectives.

618542 Is Your Plant at Risk from Climate Change?

Abstract:

Climate change is occurring and some of the changes will impact the safety of process plants. This paper summarizes potential adverse impacts on process plants and then outlines how to conduct a Climate Risk Vulnerability Assessment (CRVA) for a process plant. One product of the CRVA is a plan of action to eliminate or reduce adverse climate effects on the process plant.

619167 Is Emergency Response for Facilities Impacted By Climate Change?

Abstract:

Climate change essentially bring about alteration to our natural environment. This can be seen by way of changes in the weather pattern. We see this here in the Gulf region of the US, almost every year during our most infamous hurricane/storm season.

Emergency response plans and preparedness during this season is always critical. It is based on science, best practices and lessons learned. At these times, not only is the shutdown process very critical , however the start up and recovery measures are also critical..

This paper will discuss thoughts and views of the author of changes required to emergency response and preparedness plans that facilities have to account for, due to changes in the weather pattern. The author will also discuss how the risks to people and process during the recovery measures can best be minimized. Irrespective of the conditions and planning, the human lives must be preserved.

619069 Transitioning from One Explosive Fuel to the Next – Safety Considerations in the Emerging Hydrogen Economy

Abstract:

Since the turn of the century, there has been a focus on the development of the hydrogen economy, driven by the desire to reduce greenhouse gas emissions through utilization of alternative energies and cleaner fuels. With both the current and future interest in utilizing hydrogen, it is crucial to evaluate and mitigate the fire and explosion risks posed by hydrogen during storage and transfer. While many other fuels such as natural gas, propane, LNG, and petroleum also present flammability and explosive hazards, hydrogen has several unique characteristics. These include, but are not necessarily limited to, a relatively low ignition energy, low density, and wide flammability limits. These properties present unique challenges for risk management compared to other flammable liquids and gases that will need to be addressed as the hydrogen economy further expands into its emerging market.

In this paper, the properties of hydrogen and how they impact the fire and explosion hazard compared to other fuels will be analyzed. Then, best practices and risk management will be discussed for storage and transferring fuels, with a focus on highlighting differences that are important considerations for hydrogen. Finally, a case study will be presented that reviews an incident involving a hydrogen leak and ensuing jet fire that ignited during the removal of a portable tube trailer. The events that led to the unintended release and ignition of pressurized hydrogen, and the lessons learned from the incident will be discussed. The case study will highlight several aspects related to the safe storage and transfer of hydrogen in commercial applications.

619275 Application of Causal Analysis Based on System Theory (CAST) to the T2 Laboratories Incident

Abstract:

Application of Causal Analysis based on System Theory (CAST) to the T2 Laboratories incident

Joshiba Ariamuthu Venkidasalapathy[a,b,*]

[a]Artie McFerrin Department of Chemical Engineering, Texas A&M University, College Station, TX 77840, USA
[b]Mary Kay O Connor Process Safety Center, Texas A&M University, College Station, TX 77840, USA

Presenter email: joshvenkat03@tamu.edu

Abstract

STAMP (Systems-Theoretic Accident Model and Processes) is an accident causalty model that is based on systems thinking and systems theory. STAMP can be applied for incident investigations through the methodology called causal analysis based on system theory (CAST)[1]. It is a socio-techinial causal analysis method that systematically helps to identify causes that led to the accident. Unlike traditional accident models that rely on linear causation effects such as swiss-cheese model and domino effect models, CAST adopts a control-centric approach that helps account for causes that arise due to nonlinear interactions. Some unique features of CAST include treating safety as an emergent property, and providing the scope to account for software interations and temporal aspects of control actions. However, it is seen that applications of CAST to chemical engineering systems is scarce so far[2-4]. Considering the increasing complexity of chemical processes, accident models based on simple linear causation effects may not be sufficient in identifying the root cause of an incident.

CAST methodology is implemented in five steps: assembling the basic information, modeling the safety control structure, analysing the control structure to identify failed control elements, analysing the systemic factors and control structure flaws and providing recommendations to improve process safety. The second step, which is to model the safety control structure, is the essence of the CAST methodology. The control structure is heirarchical and requires every control block to have well defined functional objectives and control elements to achieve its goals.

In this paper, the objective is to apply CAST to a chemical incident to identify the social and technical causes that led to a chemical indident. The T2 laboratories explosion that occurred in Florida, 2007 is chosen to illustrate the CAST methodology[5]. The explosion led to four fatalities and several injuries. The CAST methodology along with extensive simulation based on MATLAB environment led to identifying both social and technical proximate causes. Most importantly, the results showed that temporal aspects of control actions, such as, how long the action was applied or how early/late it was applied also played an important role in the T2 Laboratories incident.

References

1. Thomas, J. and C. Malmquist, *Learning from Accidents that are a Consequence of Complex Systems.* Presented at the 48th Annual International Society of Air Safety Investigators, 2017.
2. Elliott, J., *CAST analysis of the Buncefield incident.* Procedia Engineering, 2017. **179**: p. 23-33.
3. Leveson, N.G., *CAST Analysis of the Shell Moerdijk Accident.* 2016, MIT. Cambridge: EU Major Accident Hazards Bureau.
4. Muñoz, M.F.T., *The Underestimated Value of Safety in Achieving Organizational Goals: Cast Analysis Off the Macondo Accident.* 2017, Massachusetts Institute of Technology, System Design and Management Program.
5. CSB, *T2 Laboratories Inc. Reactive Chemical Explosion.* 2009.

SPRING21
+17TH GCPS
A Joint AIChE and CCPS Meeting

Process Safety Manuscript Archives Available to the Practitioner

Raymond "Randy" Freeman
S&PP consulting
12303 Lake Shore Ridge
Houston, TX 77041
Rafree@Yahoo.com (for correspondence)

Ronald J. Willey, PE, PhD
Professor and Associate Chair Chemical Engineering
Northeastern University
360 Huntington Ave. 201 CN
Boston, MA 02115
R.Willey@northeastern.edu (for correspondence).

Prepared for Presentation at
American Institute of Chemical Engineers
2020 Spring Meeting and 17th Global Congress on Process Safety
Virtual
April 18 - 22, 2021

Process Safety Manuscript Archives Available to the Practitioner

Raymond "Randy" Freeman
S&PP consulting
12303 Lake Shore Ridge
Houston, TX 77041
Rafree@Yahoo.com (for correspondence)

Ronald J. Willey, PE, PhD
Professor and Associate Chair Chemical Engineering
Northeastern University
360 Huntington Ave. 201 CN
Boston, MA 02115
R.Willey@northeastern.edu (for correspondence).

ABSTRACT

Since 1967, the American Institute of Chemical Engineers (AIChE) has sponsored a series of safety symposia. The goal of these symposia is to reduce the frequency and severity of accidents in the chemical process industries (CPI). The Safety and Health Division of AIChE was established to serve as a central hub for collection and dissemination of this process safety information. Over the years, this dissemination expanded to several process safety symposia tracks at the Global Congress on Process Safety. Three symposia manuscripts have been assembled and archived into a data base accessible on the Internet through the AIChE Academy.

- Loss Prevention Symposium (LPS)
- Process Plant Safety Symposium (PPSS)
- Center for Chemical Process Safety (CCPS) Symposium

The technical papers presented at these symposia cover a wide range of topics such as fire protection, risk reduction, explosion analysis, plant maintenance practices, process safety management and many other topics.

A total of 2,524 papers were scanned and organized into a form which can be used to search for information on process safety topics. A spreadsheet of the contents of the archives was created. This paper describes the organization of the Process Safety Archives and presents an example of how the archives can be used to find information.

1. BACKGROUND

As the English philosopher George Santayana[1] (1905) eloquently stated:

"Those who cannot remember the past are condemned to repeat it"

Trevor Kletz[2] (1993) restated the importance of the failure to remember the past in his book entitled: *Lessons from Disaster: How Organizations have no memory and accidents recur.* Only people have memory. Unfortunately, our memories are very poor. Therefore, to preserve knowledge two conditions must be met:

- It must be written down
- It must be findable.

The ability to find information is very important. It does very little good to state that the needle you must rapidly find is located in one of many haystacks in the back of 40 acres of a field. Likewise stating that a document of interest may be in the Library of Congress or the National Archives achieves little in actually finding the document of interest. For information to be useful, it must be timely and rapidly findable.

Previously, the Safety and Health Division sponsored projects to scan and digitize technical papers presented at various meetings. The papers were loaded into a Folio database and made available to the Safety and Health Division membership. Several items are of interest:

- The last update of the Folio database was in 1995.
- The process safety databases (versions 1, 2, 3) use a proprietary FOLIO format and did not include PPSS.
- Better scan technology today than was available 25 years ago.

Why this is important and why is it useful? Many safety research programs were only presented at Loss Prevention Symposia and were never formally published. These include accident investigations, chemical stability tests, hazardous material property test results and other information. Such research reports may only be available in the printed conference proceedings. Rapid access to these reports and papers may improve designs and prevent future process safety incidents.

2. LOSS PREVENTION SYMPOSIUM (LPS)

2.1 History of LPS

The first LPS was organized and held in Houston, TX in February of 1967. A total of 23 technical papers were given on topics such as:

- Building Design Criteria
- Equipment Design
- Reactive Chemicals
- Process Safety Management

The final session of the first LPS was devoted to a review of six process incidents.

Since the first 1967 symposium, the Safety and Health Division has sponsored an annual LPS. More details on the history of the LPS may be found in papers by Davenport[3] (2006) and Freeman[4] (2016).

2.2 Number of LPS papers

In the time period from 1967 to 2005, a total of 1,324 technical papers have been presented and are included in the Process Safety Archives. The papers include many summaries of process accidents, as well as discussions of such topics as fire protection, safety instrumented system analysis, dispersion analysis, and nitrogen inerting to name but a few.

3. CENTER FOR CHEMICAL PROCESS SAFETY (CCPS)

3.1 History of CCPS

In the second half of 1984, several incidents occurred which had international impact:

- In October 1984 a lubricating oil tanker caught fire and exploded immediately after leaving San Francisco. One sailor died and a large oil spill resulted.
- In November 1984 a large fire and explosion occurred in a PEMEX storage and distribution center in Mexico City. Between 500 and 600 people were killed and houses were damaged to a distance of over 5 miles.
- In Early December 1984, water entered a storage tank of methyl isocynate (MIC) at a Union Carbide plant site in the Indian town of Bpohal. The resulting reaction of MIC and water caused in a runaway exothermic reaction. A large cloud of toxic material (MIC and others) was released killing at least 2,000 people and injuring 10s of thousands of people.

As a result of these disasters, 16 U.S. chemical companies approached AIChE with a request to help improve the process safety practices in the chemical industry. This request was the start of CCPS. Details on the early history of CCPS may be found in the paper by Berger[5] (2009).

3.2 Topics and intent of CCPS Symposia

The first CCPS conference was held in 1987. The conference was focused on the following areas:

- Hazard Identification
- Storage and Transfer of Hazardous Chemicals
- Reactive Chemicals
- Post-Release Mitigation
- Human Factors
- Vapor Cloud Modeling

All of these topics are of direct relevance to the Bhopal Incident.

3.3 Number of papers

In the time period of 1987 to 2006, a total of 736 technical papers were presented at the annual CCPS conference. These technical papers are focused on specific problems and included detailed reviews of lessons learned from accidents.

4. PROCESS PLANT SAFETY SYMPOSIUM (PPSS)

4.1 History of PPSS

In 1989, there were two major process incidents in the Houston, TX area.

- A fire/explosion at the Phillips Pasadena plant killed 23 workers
- A fire/explosion at the Arco Channelview plant killed 17 workers

In 1992, the South Texas Section of AIChE organized a technical symposium on process safety. The symposium was held in Houston, TX to minimize the travel to attend the meeting and to encourage engineers and managers from the Houston area chemical and process industries to attend. The first PPSS was a great success with attendance by over 700 people.

4.2 Topics and intent of PPSS Symposia

The focus of the PPSS is on plant operations. PPSS papers are focused on practical application of process safety programs. More detailed technical material is found in LPS and CCPS symposia series. Papers on the management and control of process safety programs are a popular topic. Details on the history of the PPSS may be found in the Congress paper by Edwards[6] (2012). Edwards summarizes the focus of the PPSS as:

"*Practical Application of Process Safety Management at the Plant Level: Ensuring the Protection of People, the Environment, Capital and Business*."

4.3 Number of papers

In the time period of 1992 to 2005, a total of 515 technical papers were presented at PPSS conferences. These technical papers are focused on specific problems and included detailed reviews of lessons learned from accidents.

5. CONTENTS OF PROCESS SAFETY ARCHIVES

Contents of the process safety archive include the following collections:

- Loss Prevention Symposium manuscripts from 1967 to 1980, originally Published annually as "Loss Prevention Symposium, Vol XX" by AIChE, New York, New

York, where XX went from 1 through 14. 326 titles across 14 volumes are in the archives.
In 1981, Loss Prevention Symposium manuscript numbering changed over to a AIChE session number, alpha character (e.g. 60b), mode of numbering and the AIChE discontinued the annual compilation of manuscripts into a single issue. Fortunately, John Davenport and others collected these papers and when this archive was announced, shared their personal collections with one of the authors where these were scanned in. In 1982, selected papers from the 1981 session forward began to appear in a new AIChE publication, initially called *Plant/Operations Progress*. In 1993 *Plant/Operations Progress* was renamed to *Process Safety Progress*. This collection of LPS manuscripts may also include both the *Plant/Operations Progress* version and the original manuscript delivered at the conference. Approximately 397 manuscripts and publications from 1981 to 1989 have been archived.

In 1990, AIChE moved to posting manuscripts onto digital storage media, and all manuscripts have been extracted from various digital media that existed at the time for the years of 1990 to 2005. Approximately 601 manuscripts exist in this part of the collection. After 2005, AIChE retained manuscripts from various Global of Process Safety on its web site AIChE Academy, and these can be accessed by going to the appropriate year and congress.

- The Center for Chemical Process Safety began congresses in 1987 and all manuscripts from 1987 to 2006 have been found, scanned, and archived in this database. Approximately, 736 manuscripts make up this part of the archive. Manuscripts after 2006 are available via the AIChE Academy.
- The final collection of manuscripts continues with Process Plant Safety Symposium, beginning 1992 and continued to 2005. 515 manuscripts make up this part of the archive.

Manuscripts published after 2005 are available on at www.aiche.org under this web link by year: https://www.aiche.org/conferences/aiche-spring-meeting-and-global-congress-on-process-safety

6. INTERNET ACCESS OF PROCESS SAFETY ARCHIVES

6.1 Overview of Access Protocol

Accessing the Process Safety Archives involves two major steps:

- Determine what you want to look at. This is accomplished by using the Master Index to the Process Safety Archives. The Master Index (Revision 0) is presented as supplementary material to this paper in the form of a Microsoft Excel spreadsheet. Details on how to use the Master Index are presented in section 7 of this paper,
- Once you know which volume of the LPS, PPSS or CCPS that the paper of interest is located, you must purchase the volume. The purchase price for a member of the Safety and Health Division is $0.00 (FREE). Purchase of the volume is via the AIChE transaction portal. The following section of this paper demonstrates the process for obtaining a paper from the Process Safety Archives.

6.2 Detailed Description of Method to Access Process Safety Archives

Important note: this example contains working links that if you are a member of the safety and health division, should take you to the images shown.

The process safety archives can be accessed at AIChE thru this web link: https://www.aiche.org/ccps/resources/safety-health-proceedings.

Once at the web site clicks "sign in" if you are a member of safety and health division (see Figure 1). The Home Page of CCPS will appear.

Re-click this web link: https://www.aiche.org/ccps/resources/safety-health-proceedings. to return to the archives.

This part of the archive is organized sequentially by year from 1976 to 2005. There are 7 subpages. Within each year is a link to the AIChE academy where the papers are actually stored. This approach requires the user to know what year and what conference the paper of interest exists. The easiest way to find out what to download is to use the Master Index as presented in the Supplementary Materials. For example, clicking on page 6 (bottom). Then scrolling, you will find two conferences as shown in Figure 3.

Click "READ MORE" for CCPS 17th conference. You reach this page (Figure 4).

Then you click TECHNICAL PAPERS and Figure 5 will appear.

Click the first title, Figure 6 will appear.

Click Checkout (this link is unique to the user and thus is not included).

This page appears, Figure 7.

Click Process my order. Figure 8 Will appear.

Next click CONTINUE and the following page appears (Figure 9).

You need to re-navigate after paying or receiving permission to access the conference papers. Click TECHNICAL PAPERS, and then click the first title again; this time a pdf file appears in the window (Figure 10).

This page now shows the pdf file of the original paper. Click on the paper abstract and a pdf display of the first page of the paper will appear (Figure 11).

Figure 11 shows the pdf file that can be read in a web reader or downloaded.

Another link, shown back in Figure 9 and highlighted in Figure 12, on the lower right hand side is a zip file download that will contain all papers presented in that conference, as well as the table of contents.

You can also download a zip file with all papers presented in that conference. Simply click on the Downloadable version and put the zip file onto your computer.

7. HOW TO USE THE PROCESS SAFETY MASTER INDEX -EXCEL FILE AN EASY WAY TO FIND A PAPER ON THE WEB SITE

7.1 The Process Safety Master Index

As the project progressed, an Excel File was assembled of year, symposium name, titles, authors, paper numbers. This file, to be available on the process safety web site, brought in two other data bases, the Process *Beacon* collection titles from 2001 to 2019 and the Hyperlinks to all *Plant/Operation Progress* and *Process Safety Progress* Papers published from 1982 to 2020, Issue 2. To the process safety professional, in need to seek past papers on the topic, this spreadsheet can proven to be very helpful. Two examples follow.

7.2 Ammonia Nitrate Example

An ammonia nitrate accident occurs (example Beirut, Lebanon, August 2020). The process safety expert desires to know all titles that have "ammonium nitrate" that are available in the archive:

Following these steps,

- Open the Process Safety Master Index in Microsoft Excel
- Initiate a Find and Replace dialog Box ("Ctrl F") – Figure 13
- Enter the search term "ammonium nitrate", and select the whole Workbook under option button to see all titles: - Figure 14
- Click Find All

The following list appears (in this case two manuscripts related to ammonium nitrate appear as the key word appears twice for each manuscript).

7.3 Ethylene Oxide Example

Let's try one more example, one that will have more results by searching on the combination "ethylene oxide". The following Figure 15 shows 19 titles appearing 7 titles from LPS, 1 title from CCPS (a duplicate appears), and 11 titles from *Process Safety Progress*.

Let's pick paper 3 in the list Ethylene Oxide Explosion at Doe Run. We see in Figure 16 that the paper was presented in the Loss Prevention Symposium in 1968.

Next arrow over to column M and click on the hyperlink provided:

Then, arrow over to column S and click on the hyperlink provided there (See Figure 18):

After clicking the hyperlink in Column S, Figure 19 should appear.

Figure 19. Paper Title Page

The Checkout link exists to purchase this paper. See the previous section for directions on how to purchase. If you have already purchased this volume, the pdf file link should appear below the authors (not shown in this figure).

8. DISPOSITION OF PAPER COPY OF THE ARCHIVES

The paper copies of the Process Safety Archives were scanned by InterActive of Houston, Texas. After scanning, the paper copies of the archives were deposited in the Linda Hall Library of Kansas City, MO for safe keeping. This step is still in progress as of March 2021.

9. ACKNOWLEDGMENTS

The creation of the Process Safety Archives was sponsored by the Safety and Health Division of AIChE. Creation of the Process Safety Archives using historical documents was a major task.

The Process Safety Archives includes over 2,000 technical papers, reports and documents. The Safety and Health Division Committee members were Randy Freeman (S&PP Consulting), Ron Willey (Northeastern University), Tracy Whipple (BP), Philip Myers (Advantage Risk Solutions), and Dan Crowl (Professor Emeritus, Michigan Technological University). Special thanks go to Jing Chen and Gennady Shafirovich from the New York AIChE staff. Scanning, initial file organization, and meta- data creation was completed by Maricela Manteca and her team at InterActive of Houston, TX.

The Proof Checking, Quality Assurance Team was organized by Ron Willey at Northeastern University. This team was instrumental in finding scanning issues, and cross checking the scanned material against the original conference programs to find missing papers. In several cases the QA/QC team identified papers that were missing and were retrieved by contacting the original paper author. This team included Zachary Kylar, Isabella Cardona Barber, Saule Bussing, Michael Nguy, Naa Monoh Odateifio, Joshua Garvey, and Eric Weinstein, as well as others.

9.1. Special Acknowledge for John Davenport

During the period of 1981 through 1989, the papers presented at the Loss Prevention Symposia were not collected by AIChE and published in a book as had been done in the earlier years. The years 1981 through 1989 are called the "Lost Years" by the officers of the Safety and Health Division. During this period, John Davenport of IRI collected the papers from each of the "Lost Years" as well as any associated published technical papers. Thanks to John Davenport, the collection of the "Lost Years" Loss Prevention Symposia papers have been digitized and are included in the Process Safety Archives. In many cases the collection of material by John Davenport contains both the technical paper presented at the LPS symposium and the subsequent paper as published in a journal by AIChE. Thank you John for a job well done!

10. REFERENCES

1. George Santayana, *The Life of Reason: The Phases of Human Progress, Volume 1 – The Life of Reason*, Dover Publications, New York, 1905. (Available on line at the Gutenburg Project, Ebook number 15000.)
2. Kletz, T., *Lessons from Disaster: How Organizations have no and memory and accidents recur*, Institute of Chemical Engineers/Gulf Publishing Co, Houston, TX, 1993.
3. Davenport, J., History of Loss prevention symposia: Forty years, 1967 -2006, *Process Safety Progress,* Volume 25, 2006, pp 268-273.
4. Freeman, R. "History of the Loss prevention symposium: The first 50 years – There are no secrets in safety," *Process Safety Progress*, Volume 35(1), pp 32-38, 2016.
5. Berger, S. *CCPS – 20 Years and more, Process Safety Progress,* Volume 25(1), 2006, pp 83-85.
6. Edwards, et al., *History of Process Plant Safety Symposium, Process Safety Progress*, Volume 31(3), 2012, pp 200–202.

SUPPLEMENTARY MATERIALS

1. Master Index – Excel Spreadsheet

    Contents – LPS, CCPS, PPSS, Beacon List, Index to PSP papers

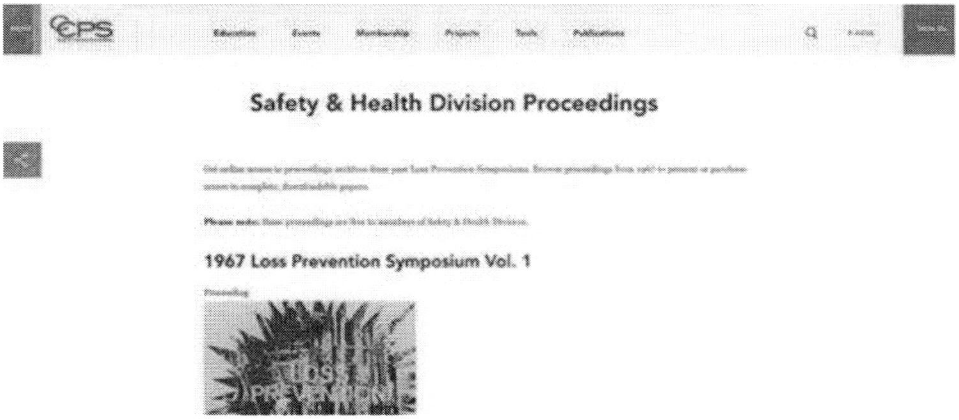

Figure 1. Initial entry page to the Safety and Health Proceedings.

Figure 2. CCPS home page.

Figure 3. Example index page that take one to the AIChE Academy Web Page.

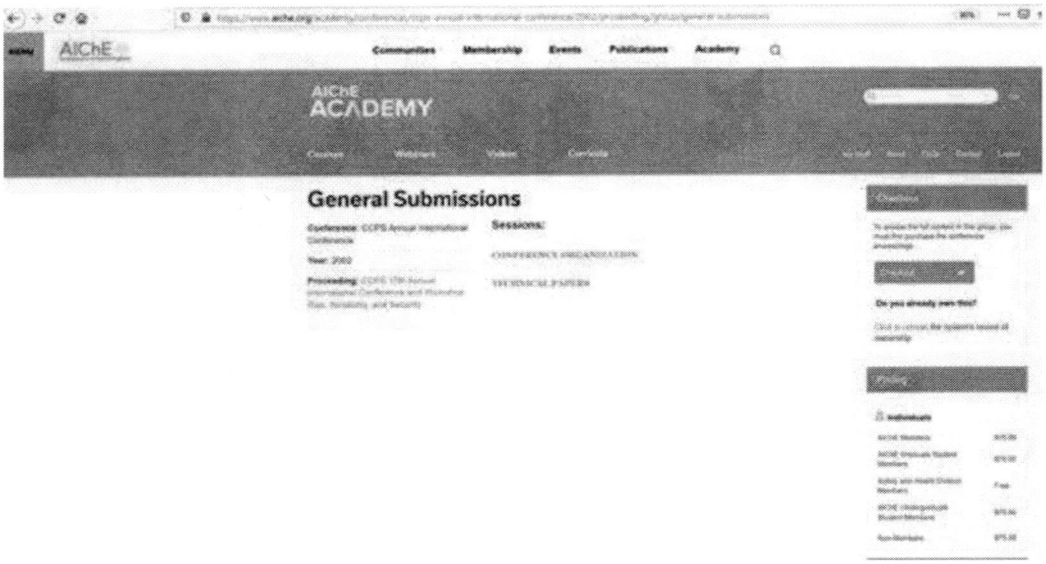

Figure 4. How to reach technical papers.

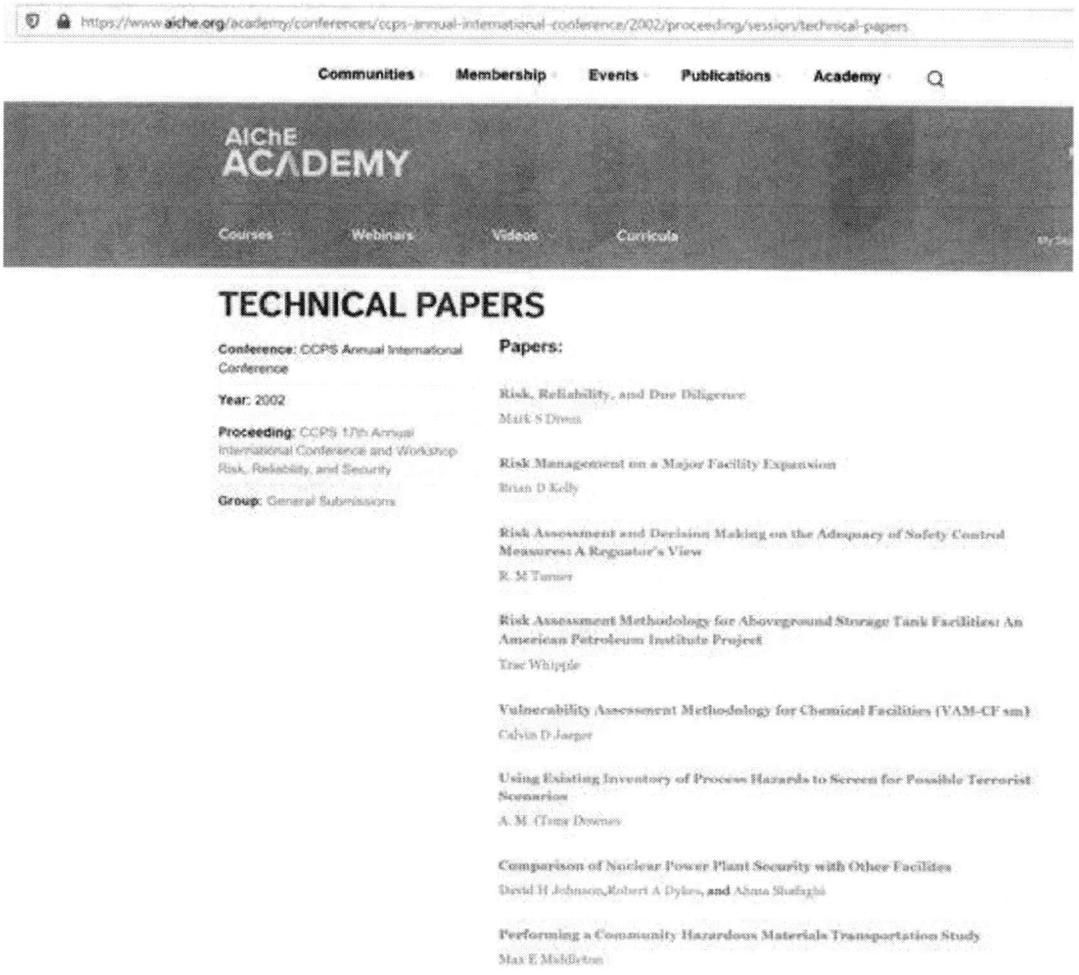

Figure 5 List of papers available for the CCPS Annual International Conference, 2002.

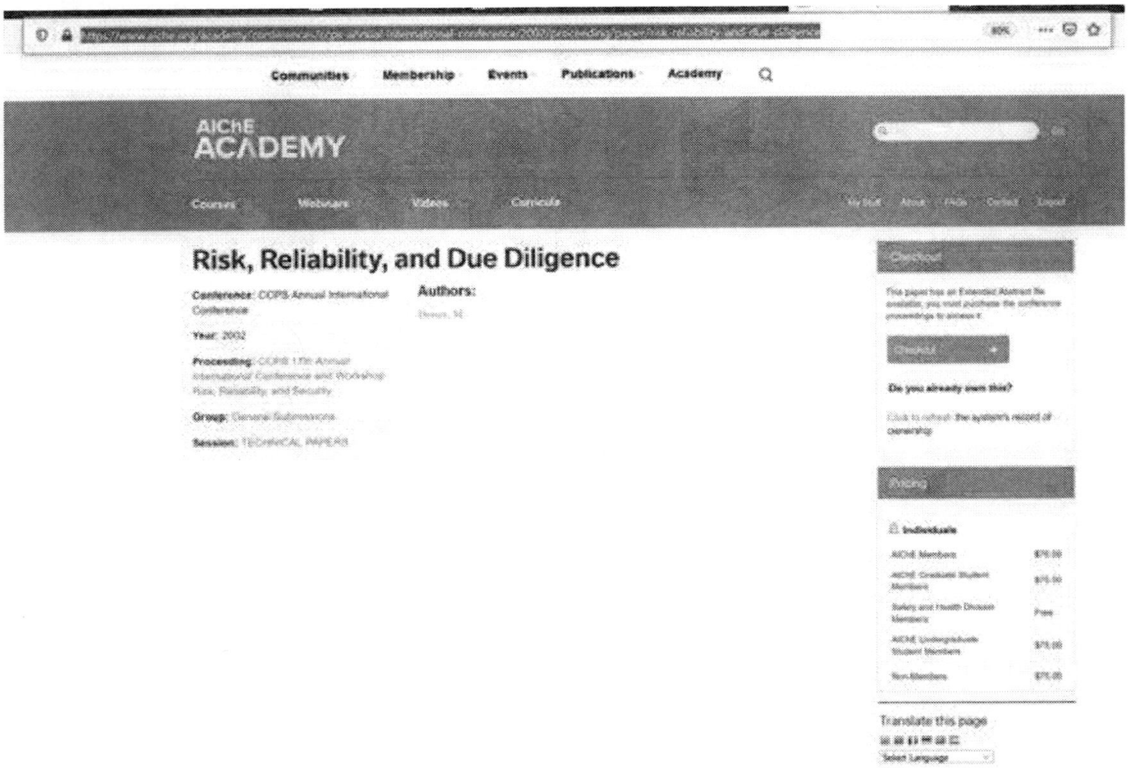

Figure 6 AIChE Academy representation of Risk, Reliability, and Due Diligence, by M. Dreux.

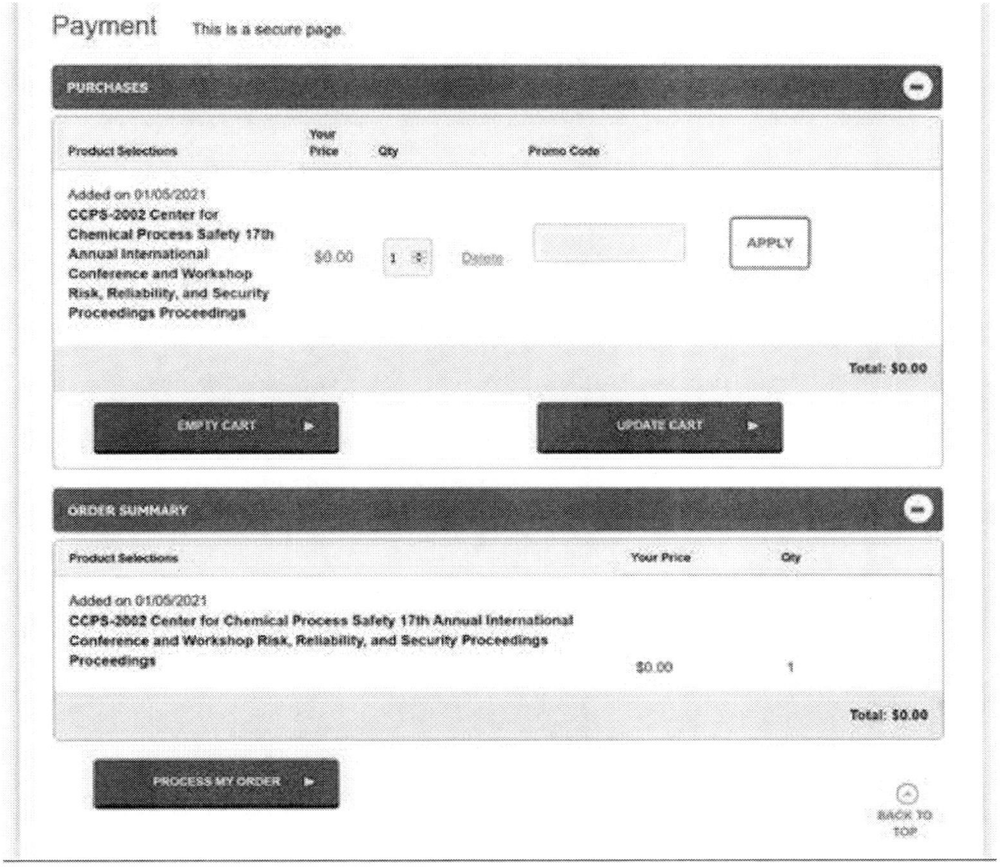

Figure 7. Purchase confirmation page.

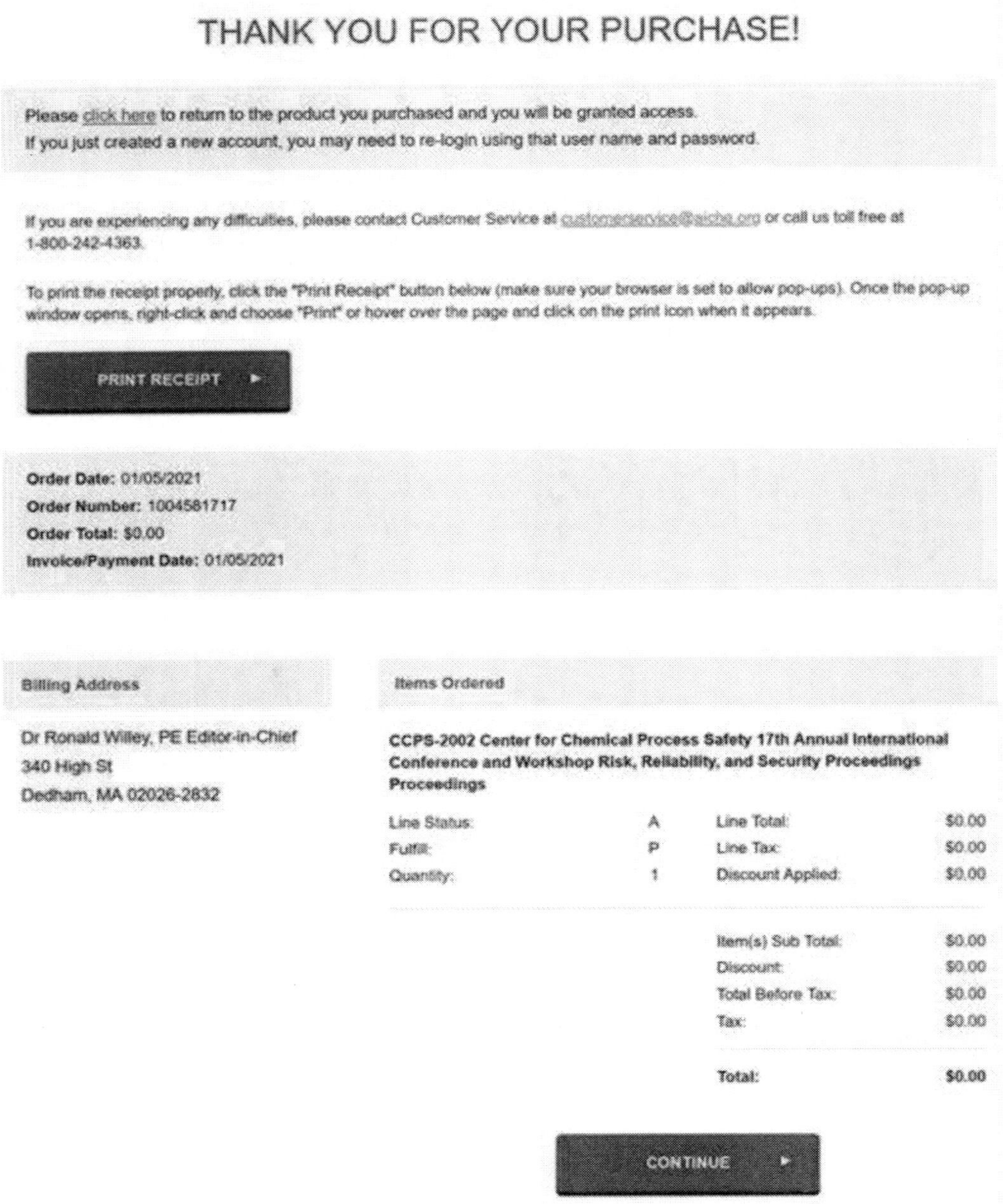

Figure 8. Receipt page

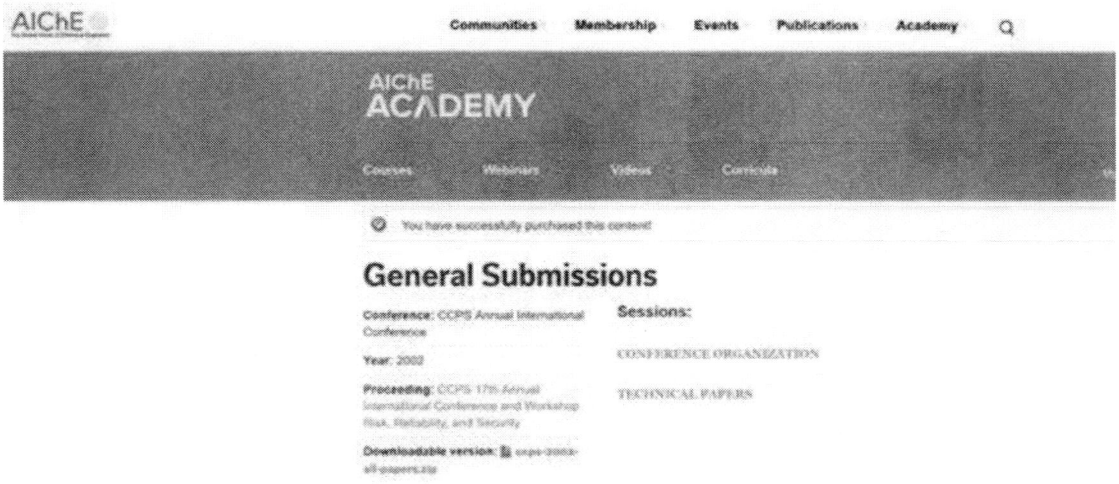

Figure 9. New View of Conference Proceedings

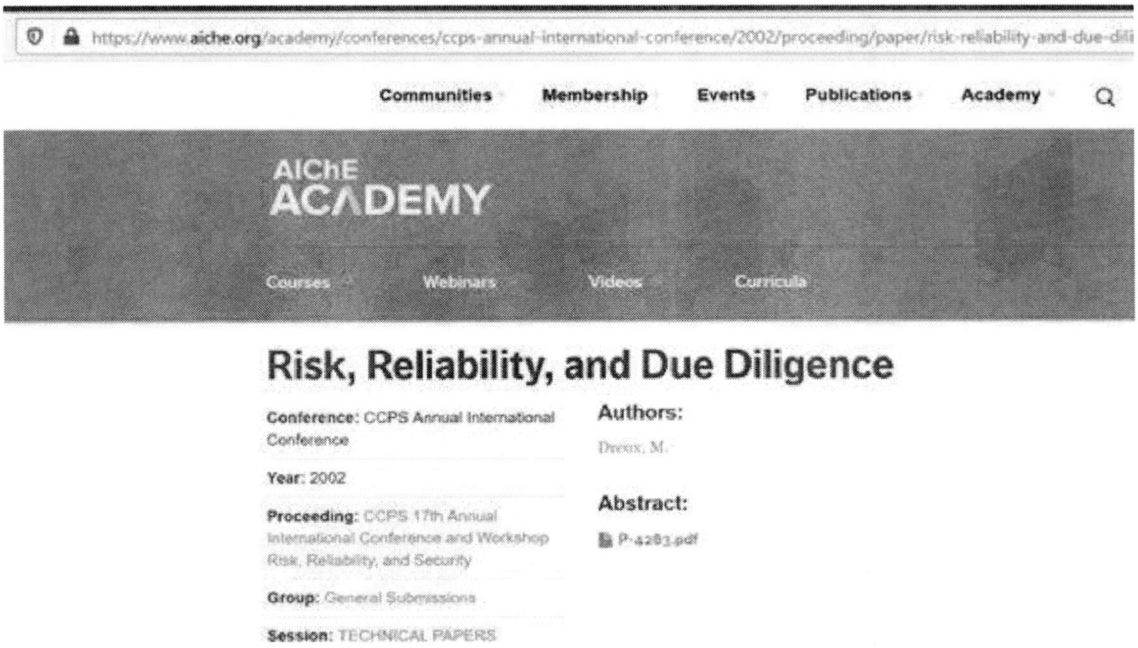

Figure 10. Paper Title Page.

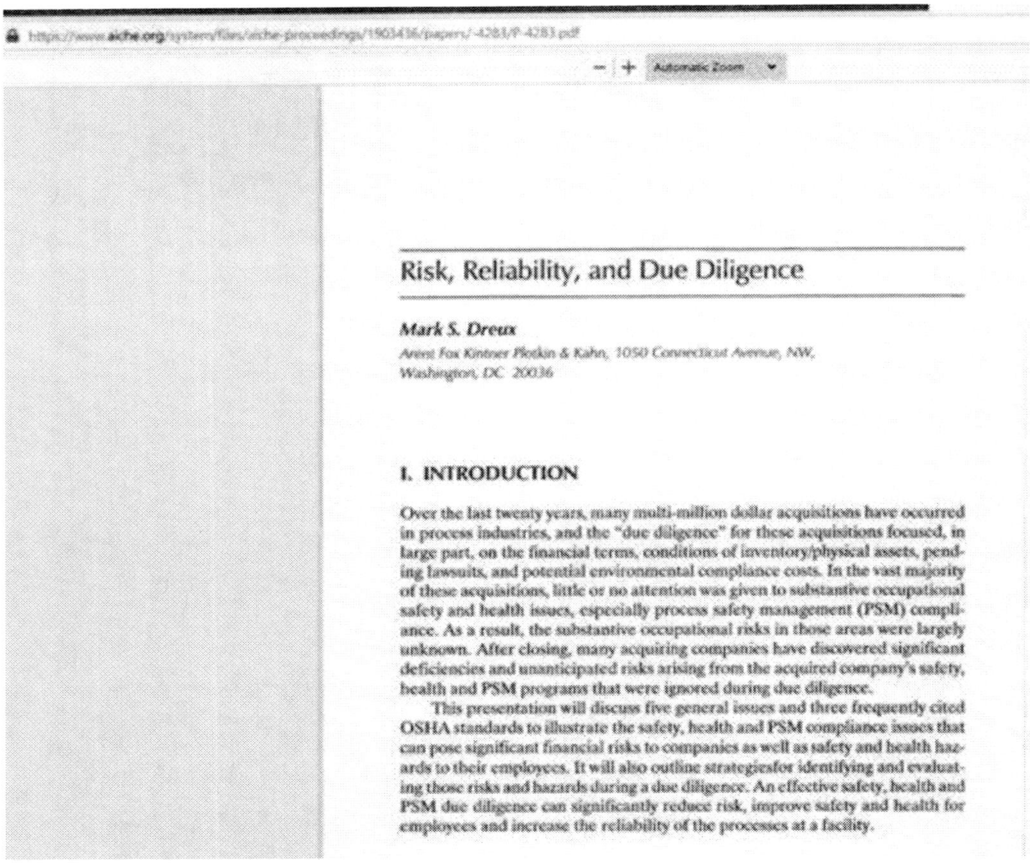

Risk, Reliability, and Due Diligence

Mark S. Dreux
Arent Fox Kintner Plotkin & Kahn, 1050 Connecticut Avenue, NW, Washington, DC 20036

I. INTRODUCTION

Over the last twenty years, many multi-million dollar acquisitions have occurred in process industries, and the "due diligence" for these acquisitions focused, in large part, on the financial terms, conditions of inventory/physical assets, pending lawsuits, and potential environmental compliance costs. In the vast majority of these acquisitions, little or no attention was given to substantive occupational safety and health issues, especially process safety management (PSM) compliance. As a result, the substantive occupational risks in those areas were largely unknown. After closing, many acquiring companies have discovered significant deficiencies and unanticipated risks arising from the acquired company's safety, health and PSM programs that were ignored during due diligence.

This presentation will discuss five general issues and three frequently cited OSHA standards to illustrate the safety, health and PSM compliance issues that can pose significant financial risks to companies as well as safety and health hazards to their employees. It will also outline strategiesfor identifying and evaluating those risks and hazards during a due diligence. An effective safety, health and PSM due diligence can significantly reduce risk, improve safety and health for employees and increase the reliability of the processes at a facility.

Figure 11. Paper finally appears.

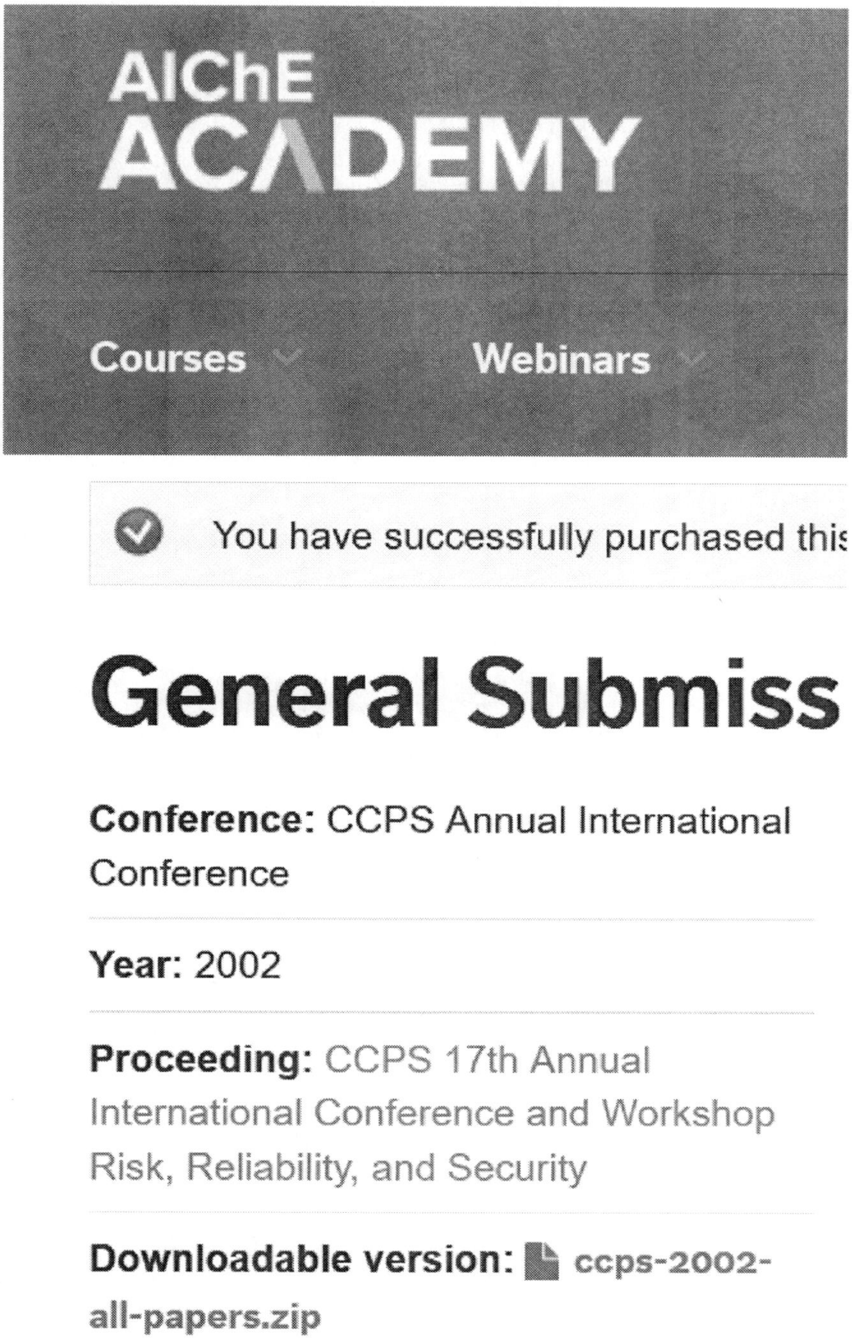

Figure 12. Conference Download Option.

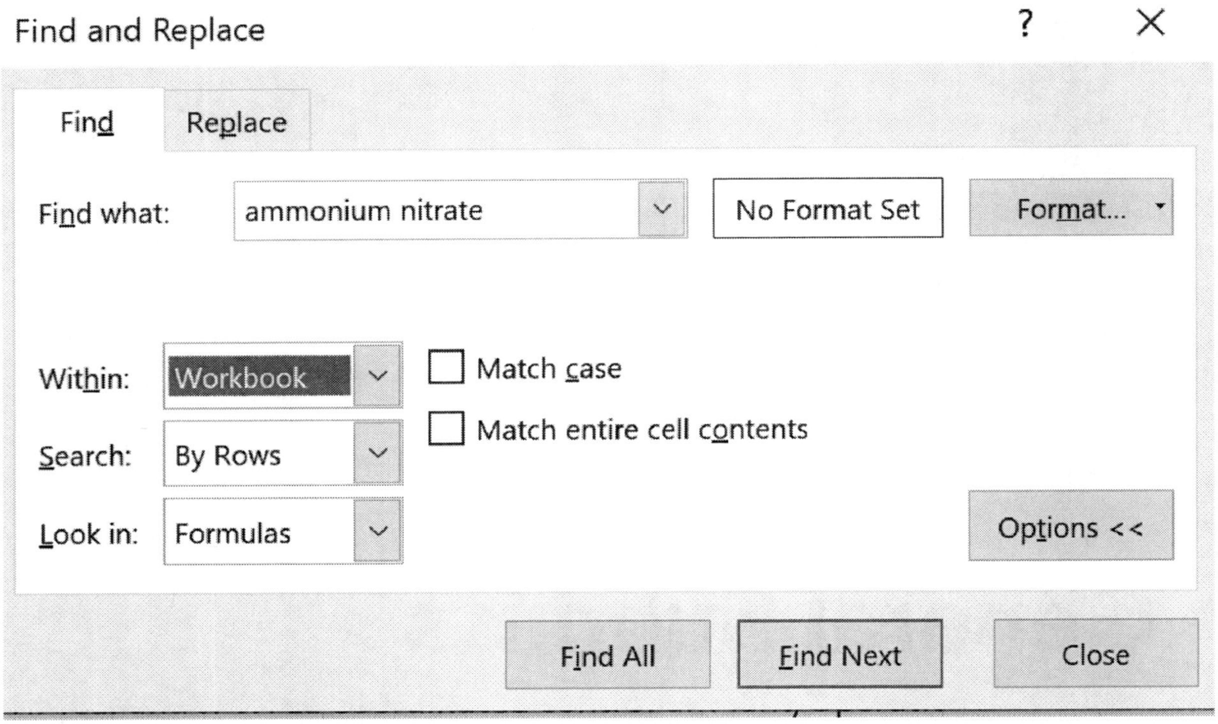

Figure 13. Setting up an Excel Search for the key word combination "ammonium nitrate".

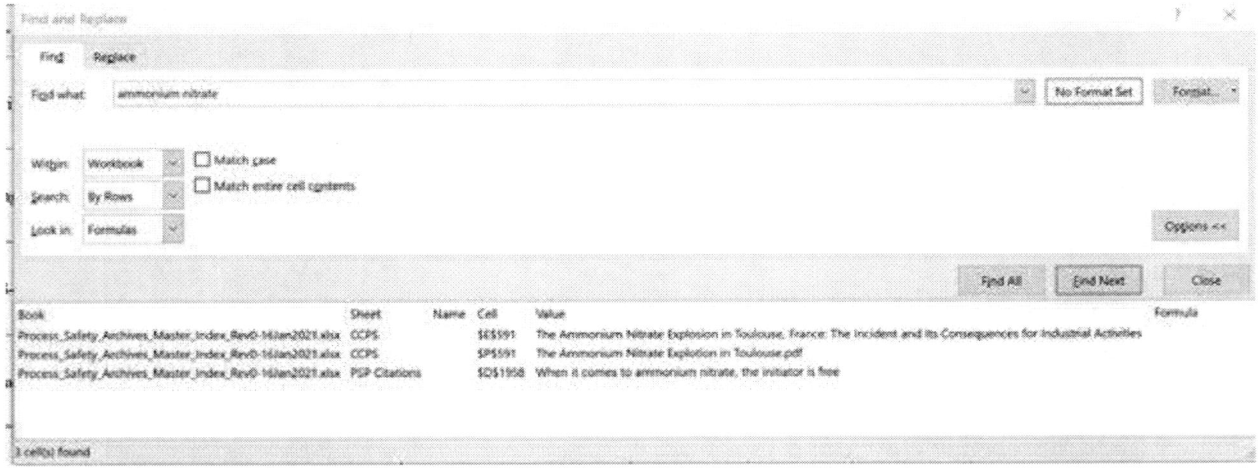

Figure 14. Search results using the term "ammonium nitrate" in the Excel Spreadsheet.

Book	Sheet	Name	Cell	Value	Formula
Process_Safety_Archives_Master_Index_Rev0-16Jan2021.xlsx	CCPS		E716	Tracing Source of Ethylene Oxide Gas House Explosion Using Damage Indicators and Thermal Modeling	
Process_Safety_Archives_Master_Index_Rev0-16Jan2021.xlsx	CCPS		P716	Tracing Source Of Ethylene Oxide.pdf	
Process_Safety_Archives_Master_Index_Rev0-16Jan2021.xlsx	LPS		E51	Ethylene Oxide Explosion at Doe Run	
Process_Safety_Archives_Master_Index_Rev0-16Jan2021.xlsx	LPS		E292	Flammability of Ethylene Oxide in Sterilizer Operations	
Process_Safety_Archives_Master_Index_Rev0-16Jan2021.xlsx	LPS		E293	Liquid Phase Decomposition of Ethylene Oxide	
Process_Safety_Archives_Master_Index_Rev0-16Jan2021.xlsx	LPS		E294	Methane Diluent Requirements for Ethylene Oxide Storage and Handling	
Process_Safety_Archives_Master_Index_Rev0-16Jan2021.xlsx	LPS		E852	Ethylene Oxide Explosion at Seadrift, Texas - Reboiler Safety	
Process_Safety_Archives_Master_Index_Rev0-16Jan2021.xlsx	LPS		E855	Ethylene Oxide Explosion at Seadrift, Texas - Iron Oxide Chemistry	
Process_Safety_Archives_Master_Index_Rev0-16Jan2021.xlsx	LPS		E863	Ethylene Oxide Explosion at Seadrift, Texas - Background and Technical Findings	
Process_Safety_Archives_Master_Index_Rev0-16Jan2021.xlsx	PSP Citations		D187	Flammability of gaseous mixtures of ethylene oxide, nitrogen, and air	
Process_Safety_Archives_Master_Index_Rev0-16Jan2021.xlsx	PSP Citations		D494	Explosive decomposition of ethylene oxide	
Process_Safety_Archives_Master_Index_Rev0-16Jan2021.xlsx	PSP Citations		D495	Thermal stability and deflagration of ethylene oxide	
Process_Safety_Archives_Master_Index_Rev0-16Jan2021.xlsx	PSP Citations		D496	An overview of exposures and exposure control in the ethylene oxide producer and ethoxylation industries	
Process_Safety_Archives_Master_Index_Rev0-16Jan2021.xlsx	PSP Citations		D497	Safe storage of dilute ethylene oxide mixtures in water	
Process_Safety_Archives_Master_Index_Rev0-16Jan2021.xlsx	PSP Citations		D500	Prediction of decomposition limits for ethylene oxide-nitrogen vapor mixtures	
Process_Safety_Archives_Master_Index_Rev0-16Jan2021.xlsx	PSP Citations		D501	Engineering controls to minimize worker exposure to ethylene oxide at sterilization facilities	
Process_Safety_Archives_Master_Index_Rev0-16Jan2021.xlsx	PSP Citations		D997	Deflagration and detonation of ethylene oxide vapor in pipelines	
Process_Safety_Archives_Master_Index_Rev0-16Jan2021.xlsx	PSP Citations		D1008	Use of propylene oxide versus nitrogen as an ethylene oxide diluent	
Process_Safety_Archives_Master_Index_Rev0-16Jan2021.xlsx	PSP Citations		D1049	Kinetics of the reactions of ethylene oxide with water and ethylene glycols	
Process_Safety_Archives_Master_Index_Rev0-16Jan2021.xlsx	PSP Citations		D1324	Analysis of ethylene oxide gas house explosion	

20 cell(s) found

Figure 15. Results using "ethylene oxide" as a search term in the Excel Spread Sheet.

	A	E	F	G	H	I	J	K	L	M
1	Master Order	Paper Title	Authors	PDF Volume	PDF Issue	Page Number	Session	Paper Location	Conference Long Title	Link to AIChE online Volume
51	50	Ethylene Oxide Explosion at Doe Run	J.E. Troyan and R.Y. LeVine	2		125-130	3	24	Loss Prevention Symposium - Vol 2	[illegible]

Figure 16. Clicking on the third entry in Figure 15 search takes one to Cell E51 of the LPS tab.

	A	H	I	J	K	L	M	N
1	Master Order	PDF Issue	Page Number	Session	Paper Location	Conference Long Title	Link to AIChE online Volume	Citation
51	50		125-130	3	24	Loss Prevention Symposium - Vol 2	[illegible]	J.E. Troyan and R.Y. LeVine, "Ethylene Oxide Explosion at Doe Run," Loss Prevention Symposium - Vol 2, AIChE, 1968 pp. 125-130.

Figure 17. Column M contains a direct link to all proceedings presented in that conference.

	A	N	S
1	Master Order	Citation	Link to AIChE online Paper
51	50	J.E. Troyan and R.Y. LeVine, "Ethylene Oxide Explosion at Doe Run," Loss Prevention Symposium - Vol 2, AIChE, 1968 pp. 125-130.	https://www.aiche.org/node/519271/group/8981/session/116896/paper/812846

Figure 18. Column S contains a direct link to the actual paper.

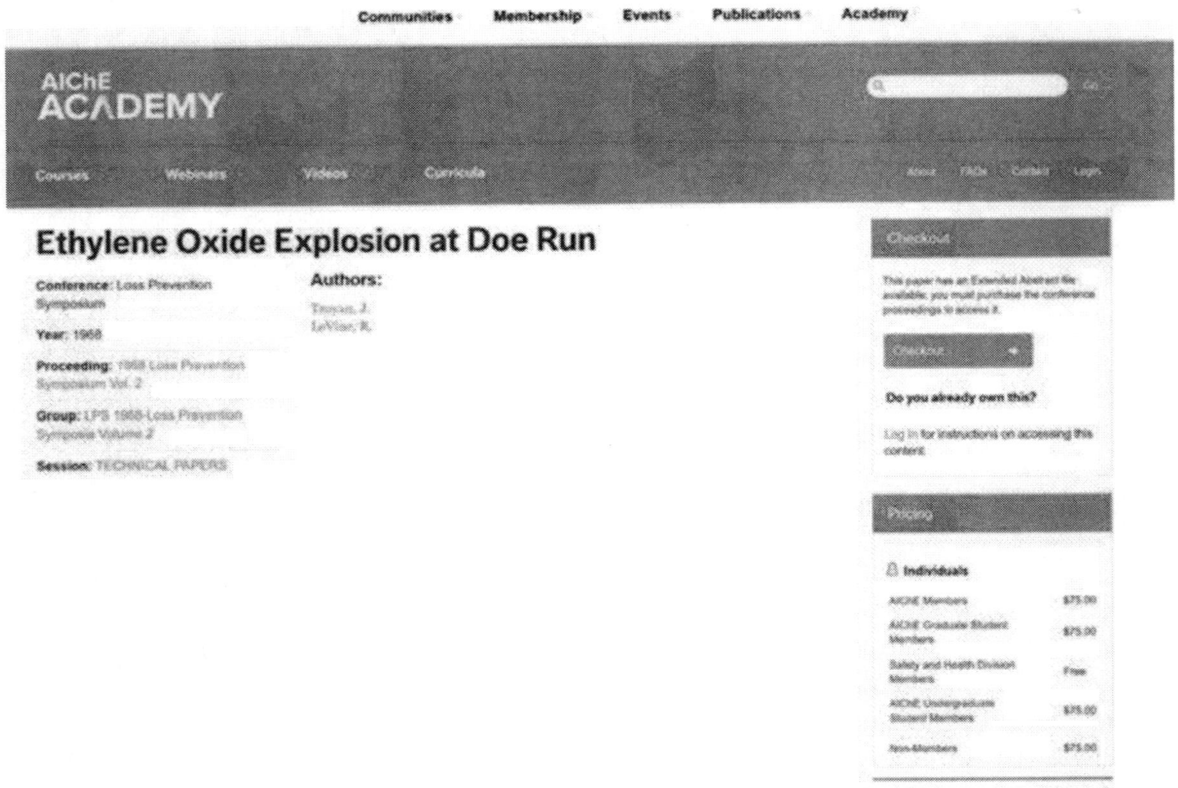

Figure 19. Paper Title Page

619027 Millennial Strategies to Capture Baby-Boomer Knowledge

Abstract:

Risk data is some of the most difficult data in the world to read, digest and understand. Focusing on Process Hazards Analysis (PHA) reports specifically, although the framework and guidewords are followed, for the most part, universally it is still difficult for someone with fresh eyes to pick up a report, read it and know what is important. Never mind having the time to wade through 500-5,000 pages of risk assessment information or multiple risk assessments looking for the specific piece of equipment that they need to know about.

Wouldn't it be great if a new operator or engineer, starts on Monday morning and by noon hour knew the most important risk information relevant to the facility or project they're working on? Why does this matter? Thinking about critical defeats processes, understanding critical procedures and which alarms matter most and need to be prioritized there are many times and processes where this information will be important.

A solution will be presented to make your risk data user friendly for millennials and all generations who are familiar with basic search engine functionality. Being able to search risk data for specific equipment tags and seeing that data in easy to understand, interactive visualizations puts it in context for what they need when they need it. Understanding the relationships between initiating events, safeguards/ IPLs (mitigative/ preventative) and consequences using bowties to tie it all together.

In addition, gamification of risk data will be presented with online, modern and interactive puzzles that can help people understand in minutes the strategies used to safeguard facilities and process units. It also adds a fun factor, competition and other nuances to engage and excite people about safety.

The future workforce needs better solutions to help enable them to leverage the knowledge of the baby boomers and generations that went before. Tens of thousands of experienced workers, who are often the SME in PHA sessions, are retiring on a weekly basis each with decades of knowledge. Although significant knowledge is being lost, the information captured in documents like PHA reports can help the next generations to learn, in second and minutes, what others took a lifetime to build. This knowledge transfer is only possible when presented in an accessible and consumable format.

619104 Lessons Learned for Nextgen PHA Facilitators

Abstract:

There are many training sessions available to train the next generation of PHA facilitators on how to execute a PHA; however, as seasoned facilitators look back on their experiences, they realize there are many details that did not get transferred in their classroom sessions. They learned these details on the job through years of valuable experience and through face-to-face and virtually-led workshops. These learnings can improve the quality and efficiency of the PHA workshop, the documentation of the workshop and the PHA experience. In 2020, this knowledge transfer became even more necessary during the COVID-19 pandemic. Organizations had to move toward conducting workshops virtually while still striving to achieve the same standards needed to ensure a safe facility and satisfy OSHA PSM requirements.

In this paper, seasoned facilitators capture both the small and the significant findings they have learned in their years of PHA facilitation from the very beginning of the PHA (at the preparation phase) to the reporting and follow-up phase. These learnings touch on setting expectations in the terms of reference, maintaining consistent documentation throughout the workshop to aid in effective reporting, maintaining consistency when the team members change (which happens in longer-term workshops), increasing engagement, especially in virtual workshop settings, and increasing use of tools (such as PHA Pro) to improve consistency and efficiency throughout the documentation process.

It is our desire to pass these lessons learned to the next generation of PHA leaders to supplement their classroom learnings and help them succeed in different settings and environments.

618410 "Begin at the Beginning and Go on till You Come to the End: Then Stop" Lifecycle of a Safety Instrumented System

Abstract:

Safety Instrumented Systems have specific steps in their lifespan. Many engineers are responsible for managing these safety systems at a specific point in the lifecycle. This seminar will describe all of the phases of the SIF lifecycle, beginning at the beginning and going through to the end.

617758 Adopting an Integrated Process Maturity Risk Model Approach to Process Safety

Abstract:

Adopting an Integrated Risk Maturity Model for Process Safety

Process Safety can mean different things to different people but one thing is for sure, that when it comes to automation we need to address the three disciplines of: alarm management cybersecurity and functional safety. The reason is that there are consequences of not addressing these from a risk point of view, such as:

- the consequence of inaction due to operators being flooded with alarms;
- the consequence of compromise due to malware or viruses that disable and/or take over the Safety Instrumented System (SIS) or Basic Process Control System;
- the consequence of failing on demand due to the SIS not being maintained properly.

The key point here is that all three disciplines are managed via performance-based standards that have a lifecycle approach with the same three phases: Analysis, Design/Implementation and Operation and Maintenance. By adopting an integrated risk maturity model, companies can assess how well they are doing at implementing these three disciplines and managing the risks associated with them, to meet tolerable levels. This paper outlines an approach and methodology to help companies establish where they are from a maturity point of view, help identify where they want to get to and how to address any gaps. The paper will also present some examples of where companies have benefited from this type of approach and that it is not just large companies that can benefit but companies of all sizes within the process industries.

617249 Reimaging Management of Change

Abstract:

For many companies, Management of Change opportunities ranks among the higher number of findings and observations of all OE processes. Additional experience has found that MOC is one of the more difficult process safety management elements to implement and to get and keep it right. But how can we improve MOC, improve leadership appreciation for what the MOC process is trying to prevent and help eliminate the "get 'er done" culture to make the change and document when we have time? Reimagine an improved overall MOC system, including people, tools, competency and efficiency.

Reimaging a MOC system would include:

- MOC system(s) are more fault tolerant and resistant to circumvention or human error
- Competent, knowledgeable personnel are involved upfront to decide what changes need to be put into the MOC process
- Qualified personnel, normally independent of the MOC originator, review the request to identify any potentially adverse impacts
- Achieve better results with fewer, qualified resources
- Monitor MOC performance and efficiency in a practical way

SPRING21
+17TH GCPS
A Joint AIChE and CCPS Meeting

YES You Can: Case Study of Implementing a Center for Chemical Process Safety Risk Based Process Safety (CCPS RBPS)

Nestor Humberto Sposito
PPR Consultora
Bahia Blanca, Buenos Aires, Argentina
pprconsultora@gmail.com
Juanita Corredor
Process Safety Leader Brinsa
Cajicá, Colombia
juanita.corredor@brinsa.com.co
John Murphy
Process Safety Services
Punta Gorda, FL
hamjfm@embarqmail.com
John Diaz
Quality Director, Brinsa SA,
Cajicá, Colombia
john.diaz@brinsa.com.co

Prepared for Presentation at
American Institute of Chemical Engineers
2020 Spring Meeting and 17th Global Congress on Process Safety
Virtual
April 18 - 22, 2021

—

YES You Can: Case Study of Implementing a Center for Chemical Process Safety Risk Based Process Safety (CCPS RBPS)

Nestor Humberto Sposito
PPR Consultora
Bahia Blanca, Buenos Aires, Argentina
pprconsultora@gmail.com
Juanita Corredor
Process Safety Leader Brinsa
Cajicá, Colombia
juanita.corredor@brinsa.com.co
John Murphy
Process Safety Services
Punta Gorda, FL
hamjfm@embarqmail.com
John Diaz
Quality Director, Brinsa SA,
Cajicá, Colombia
john.diaz@brinsa.com.co

Keywords: Process Safety, RBPS,

Abstract

This paper discusses the challenges and opportunities that arise in the implementation of a risk-based process safety management system in a hazardous materials facility in the chlorine-alkali industry, starting from a relatively low level of compliance with standards, rules of good art, and a general lack of knowledge of the basic principles of process safety.

The motivation to begin the implementation of risked-based process safety (RBPS) came from an external audit requested by the company in November 2017 to evaluate its compliance with several pamphlets from the Chlorine Institute. At that time, the organization did not have a process safety sector and safety efforts were oriented towards personal safety. The final report included a long list of recommendations, each of them associated with good practice from the CI pamphlets. These recommendations were also associated with one of the 20 elements of the CCPS RBPS.

The authors discuss how the implementation plan was developed, the awareness and training of the top and middle levels of the organization's leadership to support the project, the mechanism for selecting priorities, and the process of preparing the schedule including scope, responsibility, and compliance dates.

The paper discusses examples of improvement in facilities, equipment, systems, and procedures in the operational, maintenance, engineering, etc. sectors, the incorporation of systematic and rigorous hazard identification and risk analysis techniques, and the benefits of implementation and monitoring the recommendations suggested in the studies. The strong leadership commitment and the process monitoring methodology by the company's senior management drives the implementation and gives visibility to the management system within the entire organization.

The authors also discuss the economic results of the implementation and quantify the impact of the improvements.

The pandemic unleashed in the world, altered the course of implementation of the process safety management system. The paper discusses the alternative and exceptional measures adopted to mitigate this impact and the results of those measures.

While it is true that the progress toward total implementation of RBPS is very important, it also important to understand what is missing to achieve the full implementation of the 20 elements of RBPS, the appropriate level of maturity necessary in the culture of the organization to achieve continuous improvement and what is the plan to address the challenges to reach each stage of implementation.

The authors encourage all organizations that have not yet implemented a process safety management system to understand that it can be done, even in less than ideal conditions, with very old facilities and a poor or non-existent process safety culture, if there is a strong commitment from the top leadership of the organization.

1 Introduction

A process safety management system that is robust and reliable is necessary to ensure the safe operation of a facility involving highly dangerous products and operating conditions, such as the ones found in a chlor-alkali chemical plant.

Below, there will be a brief introduction of the installation's, which was the subject of this study, most important characteristics.

Brinsa is a Colombian company, leading in salt chemistry, with 25 years of experience. They extract and refine the salt, and from this raw material, they develop the chemistry for the chlor-alkali which supplies the whole industry. The company has manufacturing facilities in Colombia, Costa Rica, and the Dominican Republic, as well as a commercial presence in Latin America.

The chemical plant has been increasing the historical monthly production record of chlorine-soda, through electrolytic cell technology, in order to achieve the separation of the sodium chloride molecule. This is cutting-edge technology that improves process efficiency as well as environmental control.

There are inherent risks involved in the chlor alkali processing because of the hazardous substances involved in the process. Some are classified as corrosive, toxic, and flammable.

The chemical plant has an installed capacity of 100 ton/day of liquid chlorine, 225 ton/day of 50% soda, 160 ton/day of hydrochloric acid, 180 ton/day of 15% sodium hypochlorite (batch production), and 260 ton/day of 5.25% sodium hypochlorite (continuous production). Additionally, there is a storage capacity for 1540 tons of sodium hydroxide, 503 tons of hydrochloric acid, 421 tons of 15% sodium hypochlorite, 302 tons of 38% sodium chloride, 327 ton of lineal sulphonic acid, and finally, and 141 ton of chlorine as a finished product.

The plant has 4 boilers that supply steam at 400 PSIG and 400 °C. Part of the steam that is produced is used in the plant's turbo generators to produce up to 8.5 MW, achieving energy self-sufficiency. There is also a sulfonation plant, which produces raw material for the manufacturing of the cleaning products, and a blowing plant that fabricates packages and lids.

The company has a total of 1400 workers, between fixed and temporary workers and 400 permanent contractors.

2 External Audit. Trigger for the Process.

In November 2017, the company's leadership, decided to conduct third party audit of the chlorine production plant, including all of its stages (membrane, compression, storage, packaging, shipping, and distribution) to access the level of the compliance to the Chlorine Institute's Pamphlets applicable to chloro alkali production.

A protocol was developed (a checklist), based on the guidelines from the Chlorine Institute and the industry's experience. The list was used as an evaluation tool during the 5-day visit to the chlor-alkali process plant.

The evaluation activities in the field included:

- A tour of the facilities
- Review of the following programs with the plant's personnel:
    - Incident reports
    - Emergency preparation and response plan
    - Standard Operating Procedures
    - Plan for personnel training and development assurance
- Review of the pertinent Chlorine Institute checklists, with the plant's personnel.
- Presentation and discussion of the CCPS Risk-Based Process Safety program to the plant's personnel and leadership.

2.1 Findings and Priority Recommendations.

Below, are itemized the most recommendation that was given, as mandatory, by the auditors:

- Develop a written program for the investigation and report of process safety incidents. (1)
- Develop a written program for the involvement of concerned parties, including information on the process dangers, the preparation and emergency plan, self-protection measures, the alarm system, medical treatment, etc.
- Develop a plan for the performance of a periodical risk identification and analysis for the high-risk process areas.
- Develop a written plan on the reliability and mechanical integrity of the plant's assets. Define the critical equipment and an inspection and maintenance plan.
- Develop and implement a Management Of Change program (MOC)
- Develop plans of electrical areas classification in accordance with international standards
- Develop a solid culture of process safety. Establish process safety as a core value
- Provide solid leadership from all levels, starting with the board of directors (lead by example)
- Improve the operative procedures management system
- Establish a Process Safety role

The audit results were presented and shared on the last day with the facility leadership team. The team showed itself very willing to implement all of the recommendations, especially those cataloged as mandatory, and to launch a project to implement a process safety management system that didn't exist at that time in the organization.

3 Awareness & Planning

The company decided to hire the services of a process safety specialist consultancy to advise and monitor the implementation of the Risk-Based Safety System

The concept of process safety was new for all of the organization levels. Initially, a broad training and awareness plan was launched on process safety topics using different courses and workshops encompassing the different organization levels.

The plan was designed to bring them the basic process safety knowledge and to achieve the commitment from all the people to the future program implementation.

Workshops were designed with different scopes and approaches in the function of the target audience, starting with the management and leadership levels.

The first workshop was targeted at leaders from different sectors of the organization. During the 3 days training, after 24 training hours was completed, all participants were summoned to demonstrate their commitment to the system by answering the following questions:

> From your role, what is your commitment to Brinsa's Process Safety implementation?
> What strategy will you use for the implementation?

Indicate which are the 5 activities you will carry out immediately to improve the Process Safety development in your area or sector

All of the resulting material was collected in a report which was subsequently raised to general management and was a significant contribution to the further planning and implementation

Additionally, at this first stage, a 4-hour workshop was developed for the top management of the company, where top management members took part, including the company's CEO. This was key for having the commitment of the highest authorities in the organization on the management system implementation.

Table 1 shows the training that was carried out covering a total of more than 2000 men-hr of training in safety process topics by the end of 2020. During 2021 it is intended to train operators and personnel in sectors not associated with operations such as contracts, human resources, and purchases through specially designed courses for these groups.

Table 1. PS Training Course

Course Name	Top Management	Directors	Senior Management	Supervisors	Operators	Total Men-hours
RBPS Introduction			168	240		408
PSM Audit		24	168	192		384
PSM Audit Refresh for Auditors		8	56	48		112
RBPS Basic - Virtual			40	32		72
Process Safety Advanced		24	336	288		648
Inspiring Leadership	12	12				24
MOC Sensitization			46	40	180	266
MOC Implementati on			46	40	180	266

Total	12	68	860	880	360	2180

At the end of the first training, an 8-hour workshop was held to determine the planning for the implementation of the 4 pillars and the 20 elements of the Process Safety management system. The managers, chiefs, and leaders who participated were divided into multidisciplinary working teams.

The consultant developed a tool that contemplated an assessment for each element the following

- Element's impact on the organization's risk level
- Element's level of development at the organization
- Resources availability

Rating Guide
Very High - The implementation will immediately have a very strong impact on the prevention and mitigation of the risks of the installation (Weighted as 4)
High - The implementation will immediately have a considerable impact on the prevention and mitigation of the risks of the installation. (Weighted as 3 points)
Medium - The implementation will immediately have a moderate impact on the prevention and mitigation of the risks of the installation. (Weighted as 2 points)
Low - The implementation will immediately have a low impact on the prevention and mitigation of the risks of the installation. (Weighted as 1 point).

Impacto del Elemento en el Nivel de Riesgo de la Organización				
Process Safety Commitment	Very	Hig	Mediu	Lo
Proces Safety Culture	X			
Standar Compliance	X			
Proces Safety Compdetencies			X	
Workforce Participation		X		
Disclosure to Stakeholders	X			
Current score count by ranked category	3	1	1	0
Percentage of current scores by ranked category	60%	20%	20%	0%
Average Score	3.40			
Understanding Hazards and Risks	Very	Hig	Mediu	Lo
Knowledge Management Process	X			
Hazard Identification and Risk Analysis	X			
Current score count by ranked category	2	0	0	0
Percentage of current scores by ranked category	100%	0%	0%	0%
Average Score	4.00			
Risk Management	Very	Hig	Mediu	Lo
Operative Procedures	X			
Safe Work Practices	X			
Asset Integrity and Reliability	X			
Contractor Management		X		
Guarantee Training and Performance	X			
Management of Change	X			
Operational Availability	X			
Conduct in Operations			X	
Emergency Management	X			
Current score count by ranked category	7	1	1	0
Percentage of current scores by ranked category	78%	11%	11%	0%
Average Score	3.67			
Learn from Experience	Very	Hig	Mediu	Lo
Incident Investigation	X			
Measurements and Metrics		X		
Audit		X		
Management Review and Continuous Improvement	X			
Current score count by ranked category	2	2	0	0
Percentage of current scores by ranked category	50%	50%	0%	0%
Average Score	3.50			

Figure 1. Element prioritization for PSM

The teams had to classify each elements:

- High
- Medium
- Low

Then the tool performed a weighted calculation to determining the implementation priority of the each element of the management system.This was done to achieve the greatest possible impact in reducing the risk of the installation, to assure short-term implementation progress results and to maintain a very strong commitment of the organization. The workshop results and recommendations were presented to the company's management team and finally, after a few requested changes, it was approved. The plan contemplated 3 stages with priorities 1, 2, and 3 for the implementation of the different elements. The elements Process Safety Culture, Management Review, and Continuous Improvement (Stage 0) they would start in stage 1 and continue to develop during the entire implementation process. It was estimated that all the stages would be completed in 3 years. Subsequently, based on the experience, the maturation of the program, and the impact of the pandemic, it was updated extending to 5 years.

4 Organizational Strategy for Implementation.

A Company has to strive to make sure that Process Safety activities are completed in such a way that it minimizes the risk for employees, the public, and the environment, whichever the risk level. The activities that have high-risk levels could require higher attention levels from Management and higher resources assignment. The RBPS methodology intentionally guides the companies to direct efforts to the highest-risk levels sectors.

In compliance with the RBPS strategy, the organization decides to start with the Chemicals Plant, due to the high risk of the process and the results obtained from the diagnosis assessment. Likewise, a specific task force was created, in accordance with the auditor's recommendations and the industry best practices.The focus of the task force would be to develop the implementation methodologies for each of the elements of the RBPS system, follow up on the work plan, ensure interdisciplinary participation from the different sectors in the organization and ensure the system sustainability. The "Process Safety" team was created under the Quality Management integrated by a Chief, 3 professionals, and 1 analyst.

In parallel, the maintenance area team designed its preventive maintenance model and strategy. The Assets Integrity and Reliability task force, which was assigned to the Engineering and Maintenance Management, composed of 1 chief and 5 engineers. It had specific focus on preventive and predictive maintenance. The Technical Documentation center attendants were included in this work team,.They are responsible for centralizing and managing all the organization's projects and processes information.

With the emphasis on the implementation of the elements, a work plan was established with different activities, responsible and execution dates to strengthen the implementation process and to follow an effective realistic schedule.

The Process Safety Area, in alignment with the other company areas, will follow up on the progress and closure of the assigned activities. In Figure 2 a general schedule is presented of each of the elements with its implementation general plan. It should be noted that the Process Safety Culture, Management Review and Continuous Improvement elements are to be addressed the entire 5 years of implementation. Other elements started in 2018 will continue their work strategy with assistance of the consulting firm.

Elements	Start	End	% Projected to Dec/20
Process Safety Culture	02/03/2018	20/12/2022	73.30%
Management review and Continuous Improvement	02/03/2018	20/12/2022	73.30%
Contractor Management	02/04/2018	20/12/2021	73.30%
Operative Procedures	05/03/2018	30/06/2020	100.00%
Knowledge Management Process	05/03/2018	20/09/2020	100.00%
Operations Conduct	02/05/2018	20/04/2021	86.10%
Standards Compliance	01/06/2018	28/12/2021	72.10%
Hazards identification and Risks Analysis	01/06/2018	30/01/2020	100.00%
Reliability and Assets Integrity	1/10/2018	20/12/2022	
Emergency Management	01/06/2018	20/12/2019	100.00%
Training Assurance	01/06/2018	20/12/2021	72.10%
Preparation for Operation	01/07/2018	20/09/2019	100.00%
Labor Force Participation	05/03/2018	28/02/2020	100.00%
Management of Change	02/04/2018	28/12/2021	55.60%
Incidents Investigation	05/03/2018	30/06/2019	100.00%
Audit	02/04/2018	20/12/2021	71.10%
External Audit	2/11/2023	6/11/2023	0.00%
2019 Internal Audit	3/12/2018	28/02/2019	100.00%
2020 Internal Audit	20/12/2019	20/11/2020	100.00%
2021 Internal Audit	20/02/2021	20/11/2021	0.00%
2022 Internal Audit	20/02/2022	20/11/2022	0.00%
Safe Work Practices	05/03/2018	20/04/2021	89.50%
Measurements and Metrics	05/03/2018	20/12/2020	100.00%
Process Safety Competency	1/10/2018	20/12/2021	69.20%

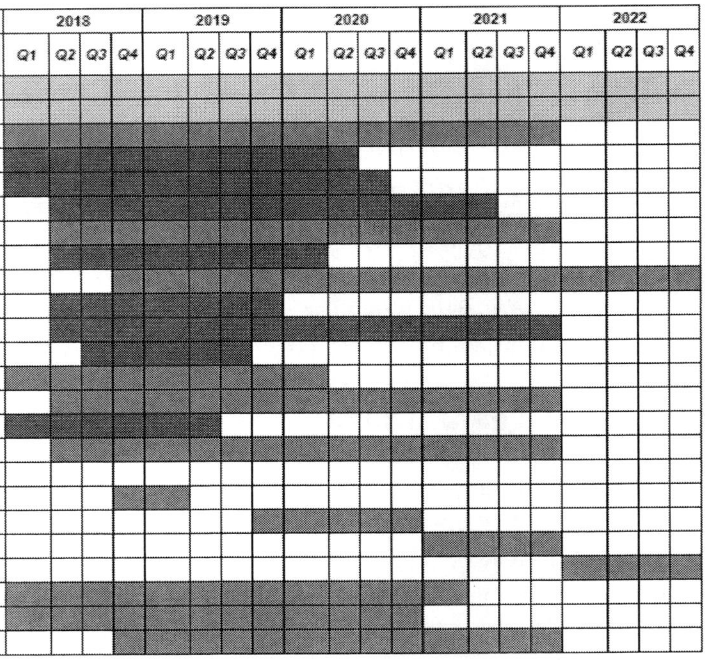

Stage 0 Stage 1 Stage 2 Stage 3

Figure 2: Management System Work Plan Implementation schedule

Next, the results of some elements of the system are presented, and the advances obtained during their development and implementation.

4.1 Hazard Identification and Risk Analysis.

The second pillar of the RBPS system "Understanding Hazards and Risks" is supported by two elements:

- Knowledge management
- Hazard identification and Risk Analysis

This pillar was identified at the diagnostic stage as High Priority(See Figure 2). Thus both elements start to be developed simultaneously, given the interdependence that exists between them. The organization decided to adopt HAZOP as an evaluation tool, complementing the company's risk matrix. HAZOP studies were developed for 10 chemical plant processes, with the participation of the entire interdisciplinary team. This made it possible to identify the current situation of the plant and assess the hazards according to the organization's risk matrix. The study was divided into two stages. Onstage implemented in 2018 included the Chlorine Circuit, Electrolysis, Hydrogen, Hydrochloric

Acid, and Sulfonation processes resulted im 162 recommendations categorized as very high and high risks The second stage started in 2019 included the CEDI (Chemical Distribution Center by its Spanish acronym meaning), Sodium Hypochlorite, Caustic Soda, Calcium Chloride and Demercurization processes resulted 176 very high and high risks recommendations.

In general, there's a high risk at the analyzed plants. 72% of the identified scenarios are at extreme or high risk. The recommendations generated by the HAZOP team when implemented significantly reduce the risk. They mainly focused on strengthening the designs with engineering measures to prevent and mitigate incidents.

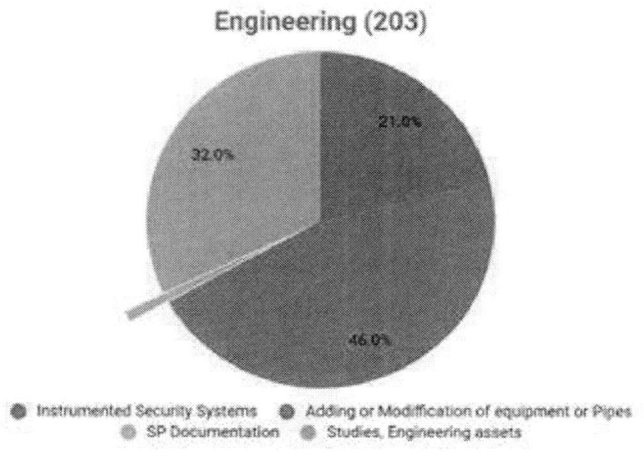

Figure 3: Engineering Recommendations

About 60% of recommendations are engineering improvements, which include among others, instrumented safety systems installation (21%), modification or adding new equipment or piping (46%), engineering studies (32%).

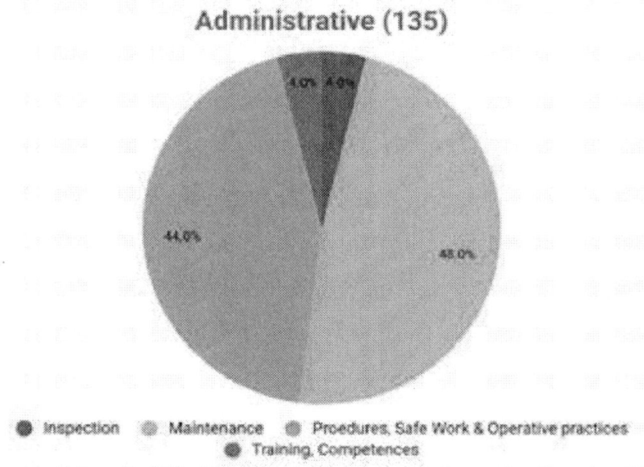

Figure 4: Administrative Recommendations

40% of remaining recommendations were administrative barriers improvements such as inspections, maintenance, training and qualification, procedures, operational and safe-work practices, and in general, process safety standards (Figure 4).

A 3-year work plan was defined for closing gaps, with budgetary assignments and defining specific task forces for gap closing.

4.2 Operating Procedures and Performance Assurance.

The RBPS element that ensures adequate development, timely maintenance, and the consistent use of operative procedures (procedures) is one of the 9 elements at the RBPS pillars: Risk Management.

Good operative procedures, among other things, describe the process, the hazards, tools, protective equipment, control systems and also provides troubleshooting instructions when a non-expected behavior emerges. They also address abnormal situations such as emergency shutdowns and temporary operations.

Operative procedures at Brinsa didn't comply with many of the requirements outlined above. Therefore, a standard is generated for drafting operative procedures according to the CCPS recommendations. This way, the participation of the operators in drafting procedures, created synergy within the personnel expertise and the information related to the operation manuals, the review by the supervision, and the coordination and approval by the Plant management. This procedure ensures that the input of different management levels is guaranteed and the documentation is appropriate. The drafting of all operative procedures required about 2 years. The final result was much more user friendly procedures, with adequate language for easy comprehension by users. Pictures were also included and unambiguous instructions of the steps to follow during the operation.

By the end of 2019, after the information and procedures updated and approved for the processes of Compressors, liquefaction, drying and Chlorine storage, de-chlorination and brine treatment, sodium hypochlorite production, calcium chloride and hydrochloric acid, a personnel retraining program was implemented. It included the technical skills assessment of the operators. it included 3 parts:

- Performance list: It is an evaluation of the knowledge of operational practices in the field.
- Theoretical evaluation of the process: The questions are asked taking into account the operating procedures.
- Cause-Effect evaluation: It measures the ability of the operator to understand the main risks that can be produced by the process and his role to minimize, delete or correct the risk.

After completing the first evaluation, feedback on weaknesses was carried out and subsequently, training was given to each operator. Finally, a second evaluation was carried out, emphasizing the non-conformities. When the training was completed all of the operators had a rating above 90%.

After updating the operative procedures, the first phase of generating the Pre-startup Safety Review (PSSR) lists was started, with the same creation, review and approval flow. In this first phase, the lists only contemplate start-ups after programmed stoppage or non-programmed stoppages without changes or modifications of the process were included. The documentation was developed during 2020. A second phase is planned for 2021that will ensure that the RFSR lists will be also mandatory for new projects.

4.3 Incident Investigation.

Developing, sustaining, and enhancing the organization's incident investigation competency is one of four elements in the RBPS pillar of ***learning from experience.*** (1)

In accordance with RBPS methodology, the company implemented a process for reporting, tracking and investigating incidents that include: 1. a formal process for investigating incidents, including staff provision, conducting, documenting and tracking process safety incidents & 2. Incident and incident investigation data trends to identify recurring incidents. This process also manages the resolution and documentation of recommendations generated by the investigations.

From the implementation of the process, a culture of reporting process safety incidents, including near misses, gradually began to emerge. Reports include, but are not limited to, chlorine leaks, spills (or drips) of liquid and corrosive substances, safety valve activation, fire outbreaks, among others. These events are investigated by interdisciplinary teams (mechanical, electrical maintenance, reliability, operation, process safety) through different investigation methodologies, with the use of Tap Root® being preferred in the organization.

The investigations involve the socialization of the lessons to be learned from the events that occurred at all levels in the operation, through daily operation meetings (DOM), and the lessons are left at different points of the plant for a free consultation. In Figure 5, the lesson-to-learn model currently being socialized

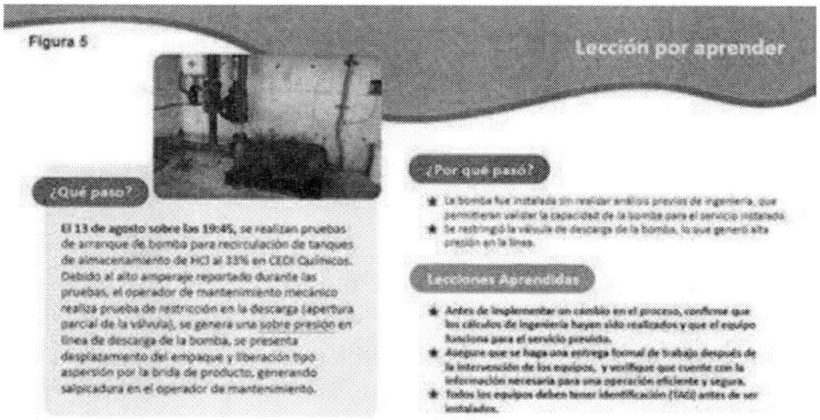

Figure 5: Lesson-to-learn model being socialized

With the implementation of this element, different levels of the organization have been reached, a culture of both reporting and interdisciplinary understanding of the events is being generated.

In Figure 6, the increase in process safety incidents reported by the Chemical Plant is evidenced. The conclusions and recommendations of these investigations led to substantial improvements in the processes, increasing safety and operational reliability.

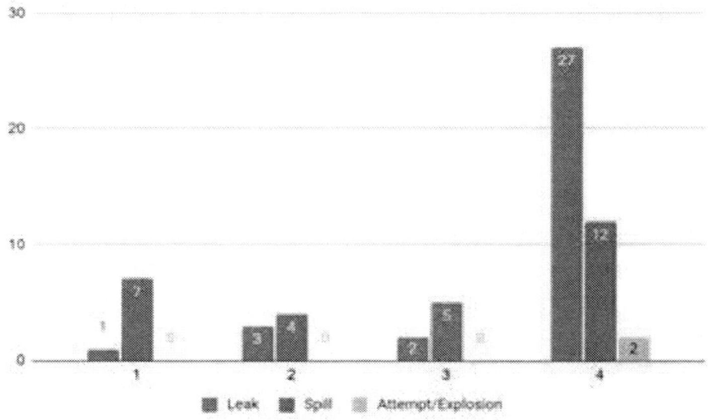

Figure 6: Process Safety reported incidents

During the first 3 years of implementation, activities of all the main elements that have allowed a systematic advance in the implementation have been developed. This has also made it possible to unite efforts from different fronts, create a comprehensive system and receive genuine support from the senior management team.

4.4Management Review and Continuous improvement

Since 2018, monthly monitoring is carried out by the management team and operations leadership to review the progress in the implementation of the process safety management system, opportunities for improvement, delays, budget execution and redefinition of strategies that leverage the system implementation. In these evaluation meetings, the gap closures identified in risk analysis and incident investigation are also reviewed. The results and conclusions of these evaluations are presented and analyzed every 4 months at the company's shareholders' meeting.

5 Investment

The RBPS implementation requires allocating human, material and financial resources to implement a robust, consistent and sustainable system over time.

The leadership of the organization, understanding its responsibility, approved for the first 4 years of implementation an approximate budget of 10,000,000 USD. This budget was assigned, among other things, to Training and performance assurance, incorporation of resources, developing the procedures and policies required by the RBPS and implementing improvements in the plant arising from the risk studies, the requirements of the management system and the conclusions of the investigation of process safety incidents.

Some improvements and changes involve large projects, and others are "minor" improvements that seek to reduce the human factor as a source of risk.

Below we will describe some of the main improvements made to the installation:

Around 30 level transmitters were installed in different process tanks, to eliminate the level glasses with which the level visualization was conducted (Figure 7).

BEFORE	AFTER
15% Sodium Hypochlorite Preparation tanks 688 / 689. Level visualization through a sight glass, without an alarm system to alert the overflow of the tanks.	Installation of level transmitters with alarms, giving the operator greater control over the operation and identification of substandard conditions, both due to overflow and lack of power.

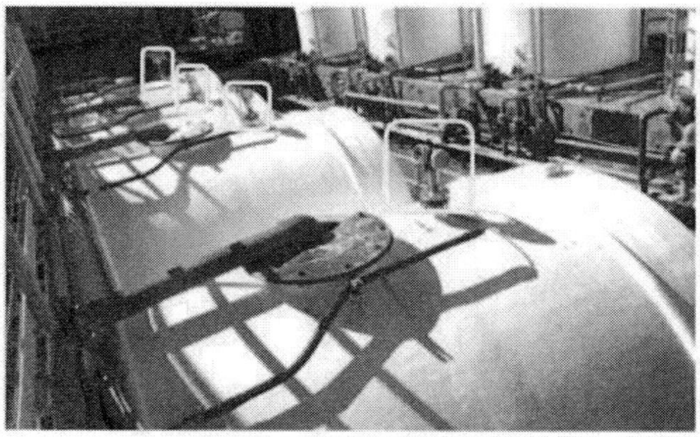

Figure 7: Level Transmitter, 15% Sodium Hypochlorite Preparation Tanks

A project was created for the installation of new liquid chlorine storage tanks, meeting the minimum requirements defined by the Chlorine Institute pamphlets, the EUROCHLOR technical definitions, and the available ASTM and ASME construction standards, in replacement of existing ones that did not comply with the RAGAGEP standards mentioned above. The storage system will start operations in the second half of 2021 (Figure 8).

Figure 8: Liquid Chlorine Storage Tanks

The implementation of the safety management system allowed the organization to know in greater depth the hazards associated with the handling of hydrogen and for this reason it was decided to carry out a specific HAZOP study for the system with the advice of a hydrogen expert. The study incorporated the risk rating using the company's matrix and based on this, approximately 40 recommendations for improvement were obtained. Most have already been implemented, especially those associated with the elimination of equipment and systems that were not operating after the change of technology from mercury cells to membrane cells.

BEFORE

Equipment in the out-of-service area of the hydrogen filtration system, excessive obstacles in the area.

AFTER

Dismantling of equipment, cleaning the area, facilitating access to operational areas.
Image was taken on May 30/20

Figure 9: Hydrogen Filtration System dismantling

In the HAZOP study carried out on the hydrogen system, it was identified that in the hydrochloric acid receiving tanks there was an imminent risk due to the presence of an explosive atmosphere (accumulation of hydrogen gas). This recommendation was verified with the technology provider, and an engineering study was started to implement the improvement. Tank blanketing, or padding with N2 was installed.

BEFORE	AFTER
Mersen Recommendation (HCl Furnace Supplier): Ensure nitrogen purge in the process tanks with minimum flow control to avoid an explosion due to hydrogen accumulation. In previous years one of the tanks had an explosion with the expulsion of part of the tank. Rec. 315. Consider adding nitrogen as a blanketing gas to these Receiver Tanks.	Connections and adjustments are made in the receiver system and gas scrubbing to maintain a permanent purge to both receiver tanks of HCl guaranteeing the displacement of remaining gases and avoiding any accumulation of hydrogen inside.

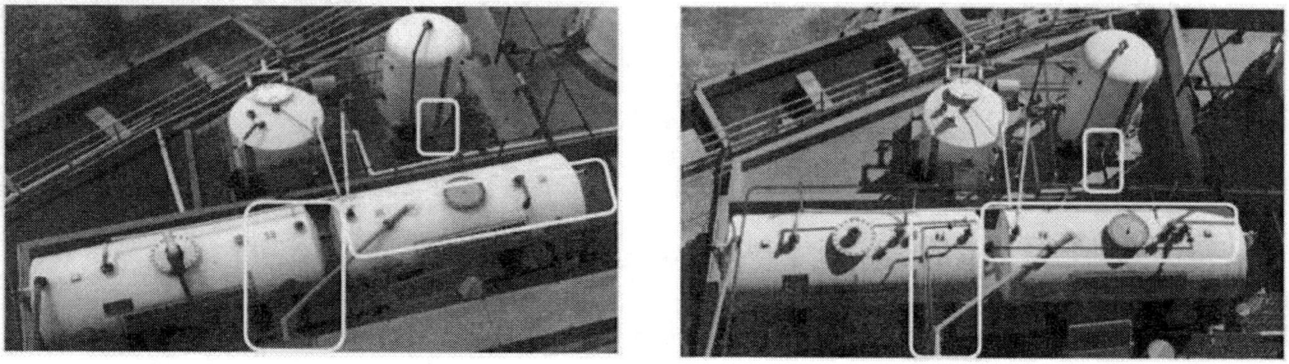

Figure 10. Blanketing gas to the hydrochloric acid receiving tanks.

Several processes in the facility still have a highly manual operation. During the HAZOP study, several recommendations were made associated with the automation of some processes. The figure shows the implementation of the automatic valve to guarantee the start-up of the backup fan of the sodium hypochlorite fan, to avoid a plant shutdown and a chlorine leak due to the pressure increase in the chlorine collector.

BEFORE	AFTER
Rec. 20. Implementation of the automatic start of the spare fan in case of fan failure in service. Risk: Loss of vacuum in the absorption tower and	Automatic valve installation and control loop programming was performed.

loss of receipt of Cl_2 with stoppage of the chemical plant due to increased pressure in the Cl_2 receipt lines and subsequent risk of Cl_2 leakage into the atmosphere with risk to people of the area.

Figure 11. Automatic startup of the sodium hypochlorite backup fan.

Consideration of the "human factor" is a key aspect of a process safety management system. This discipline takes into account, in the work environment, the interaction between people, facilities and equipment, and management systems. (2) Based on these concepts, the organization decided to incorporate them into many aspects of the management system, for example, to improve the consistency of all technical documentation, including operating procedures and PSSR checklists, which includes updating of P&IDs, datasheets, among others, a program was launched to improve the identification of all equipment, instruments and pipes in the plant. (See Figure 10, Figure 11, Figure 12).

BEFORE

Unidentified valves can lead to their misalignment and therefore a mixture of chemicals that can give off dangerous gases Cl_2, H_2S, generating risk to people in the area.

AFTER

Valves are tagged, minimizing operational errors due to confusion.

Figure 12. Valves tagging

<table>
<tr><td>

BEFORE

Tanks 769 and 779 unidentified.

</td><td>

AFTER

The identification of the auxiliary storage tanks of sodium hypochlorite at 15% is carried out.

</td></tr>
</table>

Figure 13. Tanks tagging

BEFORE

Unidentified pipe, an event occurred due to mixing of raw materials due to misalignment of valves.

AFTER

The identification of the pipes is carried out to minimize errors in the direction of fluids.

Figure 14. Pipes tagging

6 AUDIT

The only way to determine that a PSM program is working properly is to thoroughly examine both its design and implementation on a regular basis. Therefore, a PSM audit represents a way to measure the effectiveness of the PSM program and also allows a comprehensive comparison with predetermined PSM metrics. If the PSM program is not measured carefully, it cannot be controlled or improved. (3)

A successful PSM program audit begins with a commitment by the Company's senior and facility management to conduct periodic audits of these programs. This is essential to allocate appropriate resources to conduct the audits, ensure that findings and recommendations are actively and carefully addressed promptly, and ensure that on-site activities allow a successful development of the audit.

(Successful PSM program auditing according to CCPS begins with a commitment by the senior company and site management to perform periodic audits of these programs, to allocate the appropriate resources to perform the audits, to ensure that the findings and recommendations are actively and carefully addressed in a timely manner, and to ensure that the activities on-site during the audit are arranged to support the audit to the maximum extent possible.)

These principles were strongly adopted by the leadership of the organization and based on this, the process safety audits program was established to accompany the implementation of the system in order to control its effectiveness and make adjustments and corrections promptly. The plan first contemplated an internal training process and the development of audit protocols for the 20 elements of the RBPS. All this material was based on the CCPS audit book

Then it was decided to start with internal audits for partial elements of the management system. So far, 2 internal audits have been carried out, one in person in 2019 and one in 2020. Despite the challenges posed by the global situation due to the pandemic, the company decides to continue with the internal audit plan in 2020, counting with remote assistance from the consulting firm and blended participation by the audit team and the audited team, seeking to guarantee the biosafety protocols defined by Brinsa.

The audited elements in 2020 are in an advanced development process with partial implementation; a greater understanding of the spirit and scope of the audit and a more agile development were observed.

Ultimately, and beyond the qualification and findings, there is evidence of growing knowledge and commitment of most sectors of the organization with the process safety subject.

Procedures and tools have been incorporated that are strengthening the entire process safety management system. The Asset Integrity and Reliability element shows a significant improvement in its rating (2.44 in 2019 vs. 3.19 in 2020) derived from the important incorporation of procedures and tools for planning and monitoring the reliability and maintenance tasks of the equipment to be throughout their lifecycle.

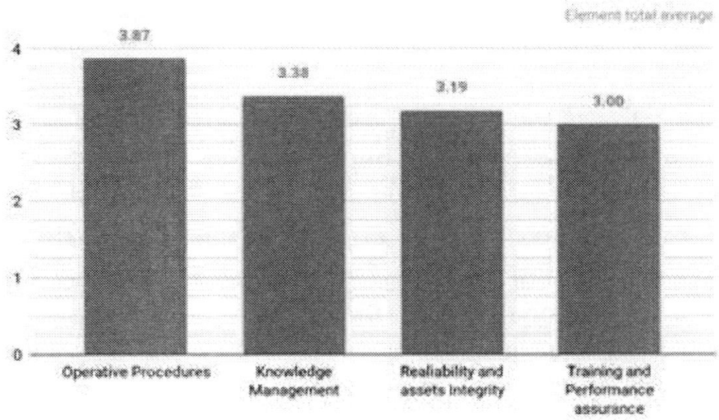

Figure 15: Qualification of audited elements 2020

The "operating procedures" element has notably improved in the second audit compared to the first one. It is worth highlighting the existence of a procedure that establishes the scope, format, responsibilities, etc. for the development of the operating procedures, and promoting the active participation of the operators in carrying out / reviewing them.

7 CONCLUSIONS

The most important conclusions of this case study are:

- An RBPS can be applied with very good results, in an organization with many years of operation and a culture focused only on personal safety.

- Awareness and training in the concepts of process safety before the implementation of the management system, at all levels of the organization, especially leadership, are essential to achieve understanding and awareness of its importance.

- The support and commitment of the organization's leadership is key to implement a process safety management system in organizations that are unaware of the issue and their safety concern is exclusively personal safety. Leadership must provide the human and material resources and make the necessary investments for a correct implementation as happened in this case.

- Implementation poses a huge challenge when starting from scratch. However, it can have very good results in the short and medium-term if it is properly planned and with the active participation of all levels of the organization in such planning.

- The proper selection of the elements of the management system on where to start the implementation is essential to try to quickly reduce the risks of the installation and at the same time show significant progress that maintains the interest and commitment of all levels of the organization, especially the leadership.

- The implementation requires a specific sector of Process Safety to lead it, monitor progress and solve limitations and challenges during the process.

- The implementation of the management system introduced benefits in other areas of the organization such as production by achieving increased reliability and operational continuity through the creation of a previously non-existent asset integrity group.

8 Bibliography

1. **Center for Chemical Process Safety, American Institute of Chemical Engineers.** *CCPS. Guidelines for Risk-Based Process Safety.* New York: Wiley, 2007.

2. **Center for Chemical Process Safety, American Institute of Chemical Engineers. CCPS. .** *Human Factor Methods for Improving Performance in the Process Industries.* New York: Wiley, 2007.

3. **Center for Chemical Process Safety, American Institute of Chemical Engineers. .** *Guidelines for Auditing Process Safety Management System.* New York: Wiley, 2011.

618537 Using Data Analytics to Generate PHA Checklists – Saving SME Time and Supporting SME Knowledge

Abstract:

The majority of those that have been exposed to Process Safety are aware of Hazard and Operability Studies (HAZOPs); a Process Hazard Analysis (PHA) methodology that is regularly used in industry because of its scalability and relative ease of execution when analyzing the risks of a process unit. Those who have been exposed to HAZOPs are also aware that these studies are qualitative, subjective and incredibly dependent on the people involved in the analysis session. Everyone who participates in a HAZOP does the best they can to do a complete and comprehensive analysis, however their contributions can be limited and biased based on their human knowledge and personal experience. This can lead to inconsistencies in identifying and capturing risks in comparable facilities/units, which can result in increased risks at site.

In order to strive towards higher quality, more comprehensive and more consistent HAZOPs, operation companies are starting to create HAZOP/PHA Checklists to help their teams better prepare for their sessions and be aware of potential high risk scenarios.

Unfortunately, HAZOP/PHA Checklists often take a large amount of time and effort for a Subject Matter Expert (SME) to create, and this amount of SME time is not always available in smaller operating companies.

So, what to do? There is obviously a benefit in having HAZOP/PHA Checklists, so how can the time of an SME required to generate one of these checklists be reduced while still creating a high quality reference tool for HAZOP teams? Also, if the insights of a few SMEs are so beneficial, what about the knowledge of ten or a hundred SMEs?

From these wants, a new method has been found to accomplish this task of generating HAZOP/PHA Checklists by leaning on the power of automatic data processing and data analysis of already completed PHAs. By combining the SME knowledge of tens to hundreds of HAZOPs of the same technology process together, a complete HAZOP/PHA Checklist can be generated with minimum manual SME input.

This paper shows a case study comparison between an SME generated HAZOP Checklist and a Big Data generated HAZOP Checklist, and highlights the high similarity/accuracy when compared, as well as the time savings of the a Big Data generated HAZOP Checklist methodology.

617639 How Effective Is Your Fire and Gas System?

Abstract:

A well-conceived fire and gas system enables the early detection of loss of containment incidents, and the resultant actions can significantly reduce the magnitude and duration of potential consequences. With a wide range of variables affecting fire and gas hazard behavior, and an unlimited number of potential fire and gas detector configurations, how does one ensure that the system is effective and fit-for-purpose?

This presentation will outline an adaptable and consistent approach to assist with the strategic development of a fire and gas detection philosophy that is applicable to both onshore and offshore environments. Emphasis is placed on the importance of hazard assessment to gain an understanding of the behavior of the specific hazards and how this can be used to inform the detection strategy, in conjunction with relevant standards and codes. The presentation goes beyond what may be included in a typical fire and gas philosophy document and covers detector layout strategies, describes modern verification methods to assess detector coverage (e.g. 3D detector mapping), and provides approaches for the management of ongoing inspection and testing of the fire and gas detection system.

618809 Factors Contributing to US Chemical Plant Process Safety Incidents from 2010 – 2020

Abstract:

Process safety incidents occur more frequently than desired – they can result in injuries, fatalities, environmental impacts, facility damage, downtime & lost production, as well as impacts on a company's and industry's reputation. While such incidents occur across many industries, this study is focused on an analysis of the most commonly reported contributing factors to process safety incidents in the US chemical industry.

The database for the study contained 79 incidents from 2010-2019, investigated by the Chemical Safety Board (CSB) and/or by companies. To be included in the database, the incident had to occur at a company classified as 325 in the North American Industry Classification System (NAICS), assigned to businesses who participate in chemical manufacturing. Unlike other summary studies of process safety incidents, the subject analysis is specific to the US chemical industry and based on contributing factors to the incidents reported post investigation.

All of the incidents included in the study were classified as Tier 1 incidents according to the American Petroleum Institute (API) RP 754 definition. Using this methodology for determining incident severity, each incident in the database included a severity score, and was placed into one of four severity categories, identified as low, medium, high and very high severity. All incidents investigated by the CSB were deemed very high severity. The list of contributing factors included in this analysis also provided more granularity and specificity than typically in other studies, and included factors such as preventive maintenance; safeguards, controls & layers of protection; and hazard awareness and identification.

Reported contributing factors were catalogued for each incident by severity category, and then a prioritized (weighted) list of contributing factors was developed for the database of incidents. The top three weighted contributing factors for the chemical industry were: design; preventive maintenance; and safeguards, controls & layers of protection. The finding that the top contributing factors to process safety incidents included factors such as operating procedures, preventative maintenance, and emergency preparedness and response, areas that are traditionally covered under the occupational health and safety administration (OSHA) regulations, prompted a review of the federal OSHA citations database. The investigation and citation history for NAICS 325 companies was analyzed to assess whether there was any overlap between the top reported contributing factors to process safety events and the top OSHA citations recorded for the industry.

A database consisting of the inspection and citation history for the chemical industry identified by NAICS code 325 was assembled for inspections occurring between 2010-2020 (August). Analysis of the thousands of entries provided an understanding of which regulations were most commonly cited, and which were most commonly cited following fatalities. Findings were presented for NAICS code 325, as well as for each sub grouping, including NAICS 3251, 3252, 3253, 3254, 3255, 3256, 3259.

The analysis of the citation history for the chemical industry specifically, identified that the key factors reported as contributing to incidents overlapped with the most common OSHA violations. The overlap of contributing factors for process safety incidents with the most common OSHA violations suggests that addressing regulatory issues may also pay dividends in reducing process safety incidents. These findings will likely be most beneficial if shared among those with the ability to create influence within the chemical industry, assess practices in these key areas, and identify improvement opportunities seeking industry best practices.

618590 Optimal Machine Learning Models for Liquid Aerosolization Safety Contributors

Abstract:

To date, many attentions have been paid to the new types of marine fuels to meet the requirement of the long-term sustainable strategy issued by the International Maritime Organization. Speaking to marine fuel selection, the safety factors should always be top prioritized. Flammable and explosive hazards, which have been well studied, are the major concerns of safety aspects of fuel application. However, the liquid aerosolization issue, making bulk liquids more hazardous on combustion and explosion, has not widely recognized in industry or academia.

This study is aimed at identifying the contributors of liquid aerosol formulation and solving their data deficiencies by developing the corresponding quantitative structure–property relationship models. 1215 liquid chemical substances and 14 predictors have been input to train the machine learning models via k-fold cross validation with the consideration of principal component analysis. Three rounds of comparisons were executed to find the final optimal models for liquid dynamic viscosity (LDV), surface tension (ST) and liquid vapor pressure (LVP). The most persuasive model for LDV is obtained by exponential Gaussian process regression (GPR) approach with seven principal components while the Matern 5/2 GPR algorithm is the most robust one for ST and LVP to formulate the optimal models. Since the reasonably good interpretation and prediction performance provided by the developed models, they can be served as effective tools to expand the database for the liquid aerosol formulation properties of organic compounds.

SPRING21
+17TH GCPS
A Joint AIChE and CCPS Meeting

Human Factors Implementation – for Plant Workers

William Bridges
President
Process Improvement Institute, Inc.
1321 Waterside Lane,
Knoxville, TN 37922 USA
wbridges@piii.com

William C. Rhodes II, P.E.
Process Improvement Institute, Inc. (PII)
82 Via Santa Maria
San Clemente, CA 92672
wrhodes@piii.com

PROCESS IMPROVEMENT INSTITUTE

Prepared for Presentation at
American Institute of Chemical Engineers
2021 Spring Meeting and 17th Global Congress on Process Safety
Virtual
April 18 - 22, 2021

Human Factors Implementation – for Plant Workers

William Bridges
President
Process Improvement Institute, Inc.
1321 Waterside Lane,
Knoxville, TN 37922 USA
wbridges@piii.com

William C. Rhodes II, P.E.
Process Improvement Institute, Inc. (PII)

Keywords: human factors, process safety, PSM, human error, risk-based process safety, RBPS, organizational culture

Abstract

Process safety is about controlling risk of failures and errors; controlling risk is primarily about reducing the risk of human error. Often it is believed that human errors committed by plant workers are the cause of most process safety accidents. However, the entire organization contributes to these human errors. Therefore, actions are needed by the organization to reduce plant worker human errors by improving Human Factors that contribute to the plant worker human errors, and to build an organizational culture that seeks to learn from plant worker human errors, rather than to assign blame.

This paper introduces the multiple types and categories of human error and the Human Factors that influence the rate at which human errors are made. It establishes the need for creating management systems for these Human Factors, and how to implement them in a manner to reduce plant worker human errors. Finally, it describes a Safety-Principled Organizational Culture that allows an organization to create the proper environment to enable these critical improvements to reduce plant worker human error. This paper builds on earlier papers, starting from 2010, on the same topic.

The data presented is from basic research by the authors on the root causes of more than 3000 accidents and near misses; and also based on the review of hundreds of accidents analyzed by others and summary data from many companies. This Video Presentation and the related slides are even more keenly focused on the selected human factors for which the frontline workers should take the lead so that the base human error rate at a site is as low as possible. Case studies and examples are used to illustrate key points.

1. Introduction

All accidents (or nearly all, if you consider that there are some natural phenomena that we either cannot guard against or choose not to guard against) result from human error. Quite often, the erroneous conclusion is drawn that the error was "caused" by the actions of the individual directly involved in the incident. However, this view not only does a disservice to the affected individual, but it prevents proper corrective actions because **<u>humans govern and accomplish all of the activities necessary to control the risk</u>** of accidents. In other words, humans influence other humans in the process. Not only do humans cause accidents (unintentionally) by making errors directly related to the process itself, but they also cause errors by creating deficiencies in the design and the implementation of management systems. Specifically, we make errors in authorities, accountabilities, procedures, feedback, proof documents, continual improvement provisions, etc. Ultimately, these management systems govern the human error rate of those directly contacting or influencing the process. Thus, the entire organization (plant workers, technical staff and management) contribute to the human errors resulting from the actions of the plant worker.

99% of accidental losses (except for natural disasters) begin with a human error ***(supported by data from more than 1500 investigations)***

Root causes of accidents are management system weaknesses
(Center for Chemical Process Safety, American Institute for Chemical Engineers, "Guidelines for Investigating Chemical Process Incidents", 2003) – OSHA agrees

Weak Management Systems → Human Error → Accidents

Figure 1: Relationship between Accidents and Management

The optimum approach to reduce plant worker human errors therefore is not to "fix" the plant worker but rather to strengthen management systems so that it is less likely that the human errors will occur (and that when they inevitably do, their consequence is controlled to the greatest extent possible).

This paper outlines this approach by first providing context to the term "human error" so that the means to reduce and mitigate it can be better understood. It then establishes that the most practical and efficient means to minimize plant worker human errors is through properly designed and implemented management systems. Next, it establishes the key Human Factors that contribute to human errors. Conceptually, plant worker human errors are controlled by understanding the Human Factors related to them and their work to the degree that effective management systems for each of them can be developed. Therefore, the paper then details how these Human Factors can be improved for plant workers. Finally, it explains that this entire process is only possible with the right type of organizational culture.

This paper introduces the concept of a Safety-Principled Organizational Culture as one that enables the proper application of Human Factors to plant workers to reduce their human errors. Further, with this culture, when human errors are made, the appropriate lessons are learned so that those errors, and even seemingly unrelated errors, are not repeated. In this way, a plant's overall safety performance will improve.

Definitions for Use in This Paper

- This paper will use the term **Human Error** to mean the errors that are made during direct interface or direct influence of the process.
- **Human Factors** are those aspects of the process and related systems that make it more likely for the human to make a mistake that in turn causes or could cause a deviation in the process or could in some indirect way lead to the increased probability of an accidental loss.
- **Management Systems** are the administrative controls an organization puts in place to manage the people and workflow related to the process under consideration, and so these inherently attempt to control human factors.

2. Types and Categories of Human Error

Quite often people will analyze a plant incident and draw the (typically correct) conclusion that it was due in large part to a plant worker human error. However, they end their analysis at that point, and are therefore unable to draw conclusions as to what could or should be done to prevent its recurrence. To fully protect against future harm, however, plant worker human errors must be fully analyzed, and the first step in this analysis is to understand the types and categories of human error.

In simplest terms, there are only two types of human error: *Errors of Omission* (someone skips a required or necessary step) and *Errors of Commission* (someone performs the step wrong). But in addition, these errors occur either inadvertently (unintentional error) or they occur because the plant worker believes his or her way is a better way (intentional error, but not intentional harm). Intentional errors can usually be thought of as errors in judgment. Some believe a "lack of awareness of the risk" causes these errors, but in actual practice, the plant worker who commits an intentional error may well be aware of the risk. They instead believe they know a better way to accomplish a task or they believe there are already too many layers of protection (so bypassing one layer will not cause any harm).

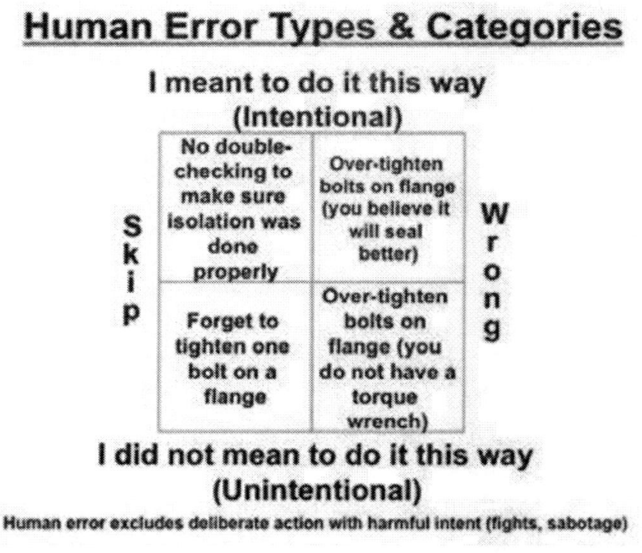

Figure 2: Types and Categories of Human Error

Regardless of type or category of human error, the organization can and should exert considerable control of the errors through properly implemented management systems.

3. Controlling Human Error through Management Systems

A process is a combination of the utilities, raw materials, and human actions (direct actions and those actions involving programming the process to accomplish automatic functions). If anything goes wrong with these Inputs, or if there are basic design flaws or basic fabrication flaws in the process, then the Outputs will not be desired. The desired output is acceptable (or high) production rates at acceptable or higher quality factors with no harm to humans (long term or short term), no harm to the environment, and with acceptable (or higher) life of the process components. The negative outcomes resulting from plant workers failing to control the raw material quality, failing to control the utility levels consistently, making errors directly related to the operation of the process, or making errors in the care of the process (such as maintenance) will result in lower production, lower quality, higher number and severity of safety-related accidents, and more negative impact on the environment. The potential of the negative outcomes is collectively referred to as ***business risk*** – more precisely, the risk is a product of the likelihood of one or more of these negative outcomes and the severity of each outcome.

A critical concept is therefore:

If an organization does not directly control risk, the organization cannot directly control quality, safety, environmental impact, or production to acceptable levels. An organization must sustainably control human error to manage the risk of accidental losses that impact quality, safety, the environment, production, or assets.

In order to sustainably control the risk of a complex process (such as an oil/gas operation, refinery, chemical plant, steel plant, automobile manufacturing, aircraft manufacturing, etc.), the organization must design and implement management systems to:

- **Understand the Risk** – This involves predicting problems; which in turn includes predicting the risk of possible accident/loss scenarios, establishing the appropriate design and the right layers of protection to control risk to a tolerable level
- **Control Risk Factors Day-to-Day** – This involves controlling the original design by maintaining the established layers of protection and managing changes to the design using integrated management systems
- **Analyze Actual Problems and Determine Weaknesses in the System** – This involves identifying weaknesses in designs and management systems and weaknesses in risk understanding through root cause analysis of actual problems (losses and near-losses)

These three elements are illustrated in the following figure:

Figure 3: Controlling Risk through Human Reliability

Management systems control the interaction of people with each other and with processes. They are the high-level procedures we use to control major activities like conducting PHAs, management of change, writing operating procedures, training employees, evaluating fitness for duty, conducting incident investigations, etc. If management systems are weak, then layers of protection will fail, and accidents will happen.

To reiterate, accidents are caused by human error. In general, Process Safety Management (PSM) is focused on maintaining these human errors at a tolerable level because:

- All accidents happen due to errors made by humans; including premature failure of equipment. There is a myriad of management systems to control these human errors and to limit their impact on safety, environment, and quality/production
- When these management systems have weaknesses, near misses occur
- When enough near misses occur, accidents/losses occur

4. Human Factors that Contribute to Human Errors

To minimize plant worker human error, process safety systems should address the **Human Factors Categories** (see various US NRC and US DOE standards from 1980s and 1990s)[1,2]. Table 1 on the next page lists the key human factor categories along with multiplication factors that poor human factors can have on the base human error rates.

TABLE 1: SUMMARY of 10 HUMAN FACTOR CATEGORIES

Based in part on: Gertman, D.; et. al., *The SPAR-H Human Reliability Analysis Method*, NUREG/CR-6883, U.S. Nuclear Regulatory Commission, Office of Nuclear Regulatory Research, Washington, DC, August 2005. PII has modified the list slightly to account for general industry data and terminology and to incorporate PII internal data.

Human Factor Category	Human Factor Issue/Level	Multiplier for Cognitive & Diagnosis Errors
Available Time	Inadequate time	P(failure)=100%
(includes staffing Issues) – ***for responses only***	Barely adequate time (≈2/3 x nominal)	10
	Nominal time (1x what is expected)	1
	Extra time (at least 2x nominal and >20 min)	0.1
	Expansive time (> 4 x nominal and > 20 min)	0.01
Stress/Stressors	Extreme (threat stress)	5
(includes staffing issues)	High (time pressures such as during a maintenance outage; issues at home, etc.)	2
	Nominal	1
Complexity & Task Design	Highly complex	5
	Moderately complex (requires more than one staff)	2
	Nominal	1
	Obvious diagnosis	0.2
Experience/Training	Low	10
	Nominal	1
	High	0.5
Procedures	Not available in the field as a reference, but should be	20
	Incomplete; missing this task or these steps	8
	Available and >90% accurate, but does not follow format rules ***(normal value for process industry)***	3
	Good, 95% accurate, follows >90% of format rules	1
	Diagnostic/symptom oriented	1
Human-Machine Interface (includes tools)	Missing/Misleading (violates populational stereotype; including round valve handle is facing away from worker)	20
	Poor or hard to find the right device; in the head calc	10
	Some unclear labels or displays	2
	Good	1
Fitness for Duty	Unfit (high fatigue level (>80 hrs/wk or >20 hr/day, no day off in 7-day period; or illness, etc.)	20
	Highly degraded fitness (high fatigue such as >15 hr/day, illness, injury, etc.)	10
	Degraded Fitness (>12 hr day and >72 hr/wk)	5
	Slight fatigue (>8 hr per day; ***normal value for process industry***)	2
	Nominal	1
Work Processes & Supervision	Poor	2
	Nominal	1
	Good	0.8
Work Environment	Extreme	5
	Good	1
Communcation	No communication or system interference/damage	10
	No standard for verbal communication rules ***(normal value for process industry)***	3
	Well implemented and practiced standard	1

If all of the human error factors shown above in Table 1 are controlled very well, then the optimized human error rates listed in Section 5 of this document are achievable. Alternatively, if one or more of these factors are compromised, the human error rate will increase by the values shown in this table. As an example, an individual whose fitness for duty rating is unfit due to the excessive work hours shown in the table will make errors at a rate 20 times greater than an individual whose fitness for duty is normal.

With excellent control of each of the human factors listed above, a company can begin to approach the lower limits that have been observed for plant worker human error. These lower limits are about:

- **1 mistake in 100 steps for most procedures-based tasks** (such as starting up a process unit), a little less for a routine (daily) task that becomes almost a reflex

- **1 in 10 chance or a little better for diagnosis and response to a critical alarm**

Excellent control requires superior design and implementation of management systems, which is enabled through a thorough understanding of these factors, as outlined below.

5. Strengthening Human Factors for Plant Workers

In order to achieve the lowest practical plant worker human error rates noted above, comprehensive management systems must be developed for each of the Human Factors provided in Table 1. Key elements for each of these Human Factors are provided below.

A. AVAILABLE TIME and STRESS / STRESSORS

The **Available Time** for the task refers to the time the task is expected to be required for completion. In many cases, the time determined for a process is determined using the theoretical times where the conditions are "ideal" and not necessarily realistic. When task and process duration are not realistic plant workers tend to "find the best and easy way" to get it done, ultimately creating latent conditions for error occurrence. **Work Stress** refers to the emotional response that arises when work demands exceed the plant worker's capacity and capability to cope. Some processes have been designed paying little attention to job design, work organization and management systems. Failing to fulfill this requirement eventually results in work stress.

Staffing levels directly impact both factors. Staffing is the process of assessing, maintaining and scheduling personnel resources to accomplish work. An adequately staffed organization ensures that personnel are available with the proper qualifications for both planned and foreseeable unplanned activities. Staffing is a dynamic process in which plant management monitors plant worker performance to ensure that overall organizational performance goals are met or exceeded. The result of an effective staffing process is a balance between personnel costs and the achievement of broader organizational goals.

Three key issues must be considered when staffing decisions are made; the first of which is selecting the right staff for a job. Each organization requires the proper amount and type of expertise to operate the plant safely and competently under a variety of conditions. The term "expertise" includes the attributes of talent, effectiveness, knowledge, skills, abilities, and experience necessary to operate and maintain plant systems, structures and components.

The second key issue is avoiding staff overload. Surges in workload, such as during outages, typically require staff augmentation as well as longer work hours for permanent staff. The introduction of contractor personnel or company personnel from other sites may increase the likelihood of errors due to unfamiliarity with the plant, its procedures and hardware, for example. Longer work hours have the potential to increase fatigue, which also contributes to the likelihood of error.

The third key factor is rotating staff every 1 hour or less for tasks that require high vigilance. Humans are inherently unable to remain alert for signals that seldom, if ever, occur. Even a sailor whose life is at stake cannot maintain an effective watch (look-out) for hostile submarines for more than 30 minutes or so. The following figure illustrates the rapid decrease of vigilance with time.

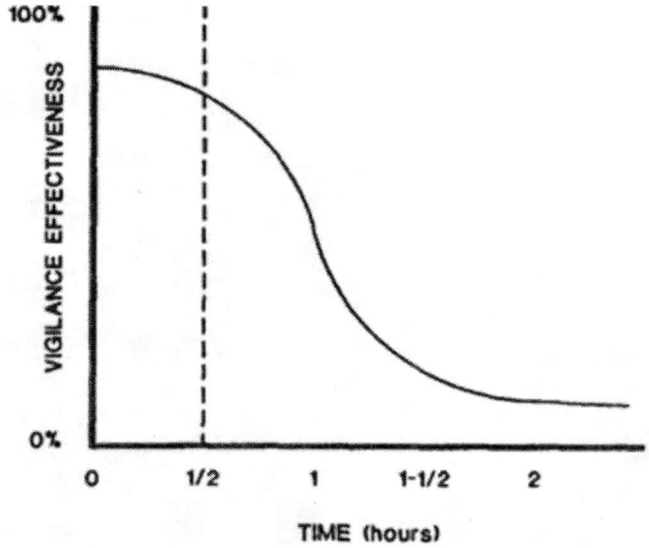

Figure 4: Vigilance as a Function of Time

Placing a worker in situations requiring extended, uneventful vigilance may lead to accidents. Therefore, it is important to design control systems in a manner that requires regular operator interaction so that the operator will remain attentive.

B. COMPLEXITY & TASK DESIGN

A task that is designed with the human limits in mind is much more likely to work effectively than one that assumes humans can and will "always" do what is written. The task must consider that humans think and remember and factor in prior data and prior experiences. Factors to consider in assessing **task design** adequacy (i.e., minimizing the likelihood of plant worker human error during its completion) include:

- Complexity of the task
- Probability of repeat errors (coupled error) on redundant aspects of the design
- Error Detection and Error Recovery

Complexity of Task (procedure-based or a call for action) – If the task is too complex, then plant workers can forget their place in the task, fail to understand the goal of each step or sub-step, or fail to notice when something isn't going right.

Task complexity is a function of:

- Number of choices available for making a wrong selection of similar items (such as number of similar switches, number of similar valves, number of similar size and shaped cans)
- Number of parallel tasks that may distract the worker from the task at hand (leading to either an initiating event or failure of a protection layer)
- Number of staff involved (more staff = more complex)
- Number of adjustments necessary to achieve the goal
- Amount of mental math required (as a rule, NO math should be required in anyone's head when accomplishing a standardized task)
- How much judgment is required to know when you have accomplished each goal within the task

For most chemical process environments, the complexity of the task is relatively low (one action per step), but for response actions (in which plant workers serve as the Independent Protection Layer (IPL)), there are almost always other tasks underway when the out-of-bounds reading or the alarm is activated. Complexity is difficult to predict (since it is not known when a human intervention will be needed), but higher complexity can increase error rates by 2 to 10 times.

Probability of Repeat Errors (Coupled errors) – For many maintenance tasks, making a repeat error or "common cause error" or "dependent error" can lead to failures of all backup systems. Such errors have led to MANY airplane crashes and major process safety accidents as well.

Coupling represents the probability of repeating an error (or repeating success) on a second identical task, given that an error was made on the first task. The increased probability of failure on subsequent tasks given that an error has already been made is known as dependence. The list below provides some starting point guidance on values to use:

- 1/20 to1/90 – if the same tasks are separated in time and if visual cues are not present to re-enforce the mistake path. This error rate assumes a baseline error rate of 1/100 with excellent human factors. If the baseline error is higher, then this rate will increase as well.
- 1/2 – if the same two tasks are performed back-to-back, and if a mistake is made on the first step of two. This error rate assumes a baseline error of 1/100 with excellent human factors. If there the baseline error is higher, then this rate will increase as well.
- 8/10 to 10/10 – if the same three tasks are performed back-to-back and strong visual cue is present (if you can clearly see the first devices you worked on), if a mistake is made on the first step of the two or more

- Two or more plant workers become the same as one person (with respect to counting of errors from the group), if people are working together for more than three days; this is due to the trust that can rapidly build.

These factors are based on the relationships provided in NUREG-1278[1] and the related definitions of weak and strong coupling provided in the training course by Swain (1993)[1] on the same topic, as shown below in Table 2.

The following relationship is for errors of omission, such as failing to reopen a root valve or failing to return a safety instrumented function (SIF) to operation, after bypassing the SIF. The qualitative values in Table 2 are based jointly on Swain[1] (1993) and Gertman (SPAR-H, 2005 which is NUREG/CR-6883)[2].

One can readily conclude that staggering of maintenance tasks for different channels of the same SIF or for related SIFs will greatly reduce the level of dependent errors. Unfortunately, most sites PII visits do not stagger the inspection, test, or calibration of redundant channels of the same SIF or of similar SIFs; the reason they cite is the cost of staggering the staff. While there is a perceived short-term higher cost, the answer may be different when lifecycle costs are analyzed.

Simple Rule: Staggering of maintenance can prevent a significant number of plant worker human errors in redundant channels. In the airline industry, the US Federal Aviation Administration (FAA) requires staggering of maintenance for aircraft with multiple engines or multiple control systems (i.e., hydraulics) (*FAA Advisory Circular 120-42A*, as part of Extended Operations (ETOPS) approval).[6]

Table 2: Guideline for Assessing Dependence for a within-SIF Set of Identical Tasks (based partially on SPAR-H, 2005 [1,2], and partially on field observations by PII) Courtesy Process Improvement Institute, Inc., All Rights Reserved

Level of Dependence	Same Person	Actions Close in time	Same Visual Frame of Reference (can see end point of prior task)	Worker Required to Write Something for Each Component
Zero (ZD)	No; the similar tasks are performed by different person/group	Either yes or no	Either yes or no	Either yes or no
Zero (ZD)	Yes	No; separated by several days	Either yes or no	Either yes or no
Low (LD)	Yes	Low; the similar tasks are performed on sequential days	No	Yes
Moderate (MD)	Yes	Moderate; the similar tasks are performed more than 4 hours apart	No	No
High (HD)	Yes	Yes; the similar tasks are performed within 2 hours	No	No
Complete (CD)	Yes	Yes; the similar tasks are performed within 2 hours	Yes	Either yes or no

Once the level of dependence is known, the probability of either repeat success or repeating errors on identical tasks can be estimated. For these probabilities, we use Table 3, which is a re- typing of Table 20-17 from NUREG-1278[1] (and the similar table in SPAR-H [Gertman, 2005][2]).

Table 3. Equations for Conditional Probabilities of Human Success or Failure on Task N, given probability of Success (x) or Failure (X) on Task N-1, for Different Levels of Dependence *Courtesy Process Improvement Institute, Inc., All Rights Reserved*

Level of Dependence	Repeating Success Equations (but shown as error probability)	Repeating Failure Equations
Zero (ZD)	$P_{Success@N} = x$	$P_{Failures@N} = X$
Low (LD)	$P_{Success@N} = (1+19x)/20$	$P_{Failures@N} = (1+19X)/20$
Moderate (MD)	$P_{Success@N} = (1+6x)/7$	$P_{Failures@N} = (1+6X)/7$
High (HD)	$P_{Success@N} = (1+x)/2$	$P_{Failures@N} = (1+X)/2$
Complete (CD)	$P_{Success@N} = 1.0$	$P_{Failures@N} = 1.0$

Error Detection and Error Recovery – Is there enough feedback in the process to allow the plant worker to realize (in time) that they made a mistake? Have they been trained on how to reason through how to recover from mistakes they or others make? (Sometimes, doing a step too late is far worse than NOT doing the step at all.) Is there enough time available for the type of intervention necessary?

C. EXPERIENCE/TRAINING

Training is necessary for general functioning of management systems and for task-specific skills such as how to start up a compressor, repair a pump, lead a root cause analysis, perform a proper Lock-Out/Tag-Out, etc. Organizations normally do a good job of training, including hands-on training. Training systems can be weak (inadequate), however, if the training system does not adequately:

- Address how to troubleshoot the process (i.e., handle process deviations or upsets).
- Provide plant workers with a mental model(s) of the process so they can perform the assigned tasks correctly, diagnose process upsets properly and quickly, and understand the consequences of their actions.
- Provide plant workers with enough practice of critical tasks.

Ironically, training is many times listed as the "cause" or "root cause" of an accident, when in fact the training is adequate, and some other Human Factor is the major cause.

D. PROCEDURES

Many procedures do not follow best practices for controlling human error, and so the written procedure actually "contributes" to increased plant worker error rates. Further, many organizations are missing guides on how to troubleshoot (what to do when process deviations occur). The best practice rules for writing and validating procedures have been published for many years ***(see Bridges, 1997-2010)***[3, 7, 8]. Below is a checklist based on the current set of best practice rules for developing operating, maintenance and other work-instructions (procedures):

TABLE 4: PROCEDURE QUALITY CHECKLIST (courtesy PII, 2008)

#	Issue	Response
	Procedure Content Checklist	
1	Is the procedure drafted by a future user of the written procedure? (Engineers should not author procedures to be used by operators or maintenance staff.)	
2	Is the procedure validated by a walk-down in the field by another future user of the procedures?	
3	Is the procedure reviewed and commented on by technical staff (engineers or vendors)?	
4	Is the procedure checked versus the Page and Step format rules below?	
5	Is a hazard review of step-by-step procedures performed to make sure there are sufficient safeguards (IPLs) against the errors that will occur eventually (when a step is skipped or performed wrong)?	
6	Is the content measured using "newly trained operators" to judge the % of steps that are missing, steps that are confusing or wrong, and steps that are out-of-sequence? (A score of 95% accuracy of content is good.)	
	Page Format Checklist	
1	Is the title of the procedure the largest item on the page?	
2	Is the procedure title clear and consistent with other titles, and does it uniquely describe the topic?	
3	Are the document control features the smallest items on the page?	
4	Are temporary procedures clearly identified?	

#	Issue	Response
5	Is white space used effectively? • Is there one blank line between each step? • Does indentation help the user keep their place? • Are the margins large enough to reduce page congestion?	
6	Is type size 12 point font or larger?	
7	Is serif type used (rather than sans-serif)?	
8	Is mixed case used for words of steps, with ALL CAPS used only for special cases (such as IF, THEN, AUTO, and WARNING)?	
9	Is the step number very simple (single level of number used)? Only an integer?	
10	Have sections or information not necessary to performing the steps been moved to the back or to another part of the manual or training guide?	
11	Are section titles bold or larger than the text font? Do sections have clear endings?	
12	Is the decision on electronic presentation versus hard copy correct? Are electronic linkages to procedures clear and accurate and easy to use? If paper is chosen, is the color of the paper appropriate?	
13	Is the overall page format (such as Outline format or T-Bar format) appropriate to the use of the procedure?	
14	Are play script features added for tasks that must be coordinated between two or more users? • Play script is normally used when there are two or more hand-offs of responsibility for steps.	
15	Are rules followed for formatting of Warnings, Cautions, and Notes? (See annotated rules, such as Warnings are for worker safety and Warnings must clearly standout from rest of page.)	
	Step Writing Checklist	
1	Is each step written as a command?	
2	Is the proper level of detail used throughout? This is judged based on: • *Who* will use the procedures • Same level of detail used in similar procedure steps	
3	On average, is there only one implied action per instruction? Best practice is to average 1.2.	
4	Does the procedure indicate when sequence is important? • If sequence matters, each step should be numbered (with an integer or letter) • If sequence does not matter, bullet lists should be used	
5	Are only common words used? Apply "education" level test (5 grade reading level is best)	
6	Do all <u>acronyms</u>, <u>abbreviations</u>, and <u>jargon</u> aid understanding? • Develop a list of such terms for use in procedures *and* communication. • Use terms that users use (within reason)	
7	Is each step <u>specific</u> enough? No room left to guess/interpret: • The meaning of a word or phrase (Check vs. Make sure) • The intent of a step or series of steps • A desired quantity or value • To what equipment the step applies	
8	Is the procedure free of steps that require in-your-head <u>calculations</u>? • Values expressed as ranges rather than targets with error bands • Conversion tables, worksheets, or graphs provided where needed	
9	Are graphics to the user's advantage? • No explanatory paragraphs or lengthy instructions that could be replaced by a picture • No impressive graphics that provide no real advantage	
10	Are references to the user's advantage? • No lengthy explanations or instructions that could be replaced by branching to a reference • No references to a procedure that references still another • No gaps or overlaps between this procedure and a referenced document • If branching, must branch to a procedure, not to a specific step in a procedure	

#	Issue	Response
11	Are rules followed for writing warnings, cautions, and conditional steps? • No more than 2 per page • No actions within a warning or caution (actions must always be numbered steps) • Warnings and Cautions contain descriptions of potential consequences	

For procedures to be effective in ensuring that tasks are performed correctly, they must be used. There are many reasons that plant workers may not use procedures. ***Deficient Procedures*** are the most prevalent problem in process industries since procedures have not traditionally been developed from the perspective of optimizing human factors; instead, procedures have been traditionally developed to meet a compliance requirement to have written procedures. Examples of procedure deficiencies (inaccuracies) include:

- Incorrect/incomplete/nonexistent. Most procedures we have audited have been only 70 – 85% accurate; the inaccuracies include missing critical steps, steps as written are not what needs to be done, or the steps are out of sequence
- No/misplaced warnings. For example, a warning should never **contain** the action to take; it should instead **emphasize** the action to take
- Poor format and presentation rules

Other reasons plant workers may not use procedures include:

- Procedures are out of date
- No procedure has been written for the task
- Plant workers cannot find the procedure they want to use
- Plant workers don't need a procedure because the task is simple
- Plant workers need more information than the procedures contain
- Plant workers see procedures as an affront to their skill
- Procedures are difficult to use in the work environment
- Procedures are difficult to understand

So, in addition to the rules for writing procedures that are shown in the table above, the organization must also address the reasons that cause the plant worker not to use the written procedure.

E. HUMAN-MACHINE INTERFACE (INCLUDING TOOLS)

The **Human-Machine Interface** (also known as the Human-System Interface, or HSI) is defined as the technology through which plant workers interact with plant systems to perform their functions and tasks. The major types of HSIs include alarms, information systems, and control systems. Each type of HSI is made up of hardware and software components that provide information displays, which are the means for user-system interaction, and controls for executing these interactions.

Plant worker use of HSIs is influenced directly by (1) the organization of HSIs into workstations (e.g., consoles and panels); (2) the arrangement of workstations and supporting equipment into facilities, such as a main control room, remote shutdown station, local control station, technical support center, and emergency operations facility; and (3) the environmental conditions in which the HSIs are used, including temperature, humidity, ventilation, illumination, and noise.

There are three important goals to be achieved in the design and implementation of the HSI. These are:

- **Design for operability** refers to designing the HSI to be consistent with the abilities and limitations of the plant workers who will be operating it. Weaknesses in the design processes can result in an HSI that is not well suited to the tasks that they must perform to ensure plant safety, resulting in increased workload, decreased performance by personnel, and an increased likelihood of errors.

  The wide-spread adoption of distributed control systems has provided many advances in process control. However, if implemented improperly, these systems can greatly reduce the operability of the facility. Although they were unable to provide as much information, old analog panels were typically laid out in a manner such that experienced plant workers were able to monitor the entire process and effectively identify conditions meriting attention. Due to space limitations, and their cost (e.g., hardware and installation), alarms were typically limited to those deemed to be associated with a significant process upset.

  The advent of distributed control systems has eliminated the space limitations associated with alarms, and greatly reduced their costs (in many cases, alarms can be added merely through programming). Without proper alarm prioritization, configuration structure and discipline, this can lead to "alarm flooding" (an excessive number of simultaneous alarms) during abnormal operating conditions. The human error of failing to respond properly to an alarm increases by a factor of over 1,000 if ten alarms activate simultaneously compared to a single alarm activation.

  Therefore, it is critical that control and alarm system designers in facilities with distributed control systems properly prioritize information so that plant workers can respond appropriately to abnormal conditions. Further, trouble-shooting guides should be readily accessible via the distributed control system to reduce human error in alarm response.

  Beyond the topic of alarms, the proper design of the visual displays can also reduce human error. Certain color combinations, such as a light gray background with darker gray equipment and black lines coupled with a flashing red indication for those components in an alarm state effectively draw the plant workers' attention to the relevant part of the process requiring attention. Further, an effective display should include:

  - Safety systems that are bypassed, impaired, or out of service
  - Process alarms that are disabled
  - Process conditions that are not within established limits
  - Backup systems that are unavailable
  - Special maintenance or testing activities that are currently in progress

- **Design for maintainability** refers to designing the HSI and associated plant equipment to ensure that plant workers are able to perform necessary maintenance activities efficiently. Weaknesses in the design process can result in systems that impose excessive demands on plant workers for maintenance and, therefore, are prone to maintenance errors or problems with reliability and availability.

- **Design for flexibility** refers to the way that changes, such as upgrades to the HSI, are planned, and put into use. A new HSI component may require the plant worker to perform functions and tasks in new ways. Skills that the plant worker developed for managing workload when using the former design, such as ways for scanning information or executing control actions, may no longer be compatible with the new design. The new HSIs must also be compatible with the remaining HSIs so that plant workers can use them together with limited possibilities for human error. Also, HSI modifications may not be installed or put into service all at one time, causing the plant worker to adapt to temporary configurations that are different from both the original and final configurations. Weaknesses in HSI implementation can increase plant worker workload and the likelihood of errors.

Tools are a special category of HSI, which typically refers to hand tools or devices that are generally designed with the plant worker in mind. A big part of human factor consideration is how to make the equipment and process operation mistake-proof (to prevent errors as much as possible). The following are some factors to consider when mistake-proofing in designs:

- **Design for unambiguous assembly.** Design the product or device such that the assembly process is unambiguous (by designing components so that they can only be assembled one way); i.e., design matching parts that are easy to insert and align. For example, use notches, asymmetrical holes, and stops to mistake-proof the assembly process. Products that go together in only one way require less plant worker training, perform more reliably, and can be repaired more quickly.
- **Consult plant workers.** Operators, technicians, and maintenance personnel can pinpoint the most troublesome areas.
- **Avoid symmetry.** When a particular orientation is critical to the design, avoid symmetry. For example, use nonsymmetrical hole patterns.
- **Use labels sparingly.** Labels tend to come off equipment too easily and often are wordy.
- **Review the environment.** Environmental problems that encourage mistakes include poor lighting; high/low heat; excess humidity, dust, and noise — anything that distracts plant workers.

The following are some factors to consider when mistake-proofing in process/operations:

- **Error-proof Mechanisms.** Error-proof mechanisms are very powerful in improving system reliability when incorporated into the design. These mechanisms, by design, will not allow a plant worker to perform an illegal operation. For example, if a plant worker enters a value that is outside the accepted range of operation, the control logic will not accept the value.

- **Automatic Alerts.** Automatic alerts immediately inform the plant worker of an illegal operation to prompt corrective action. These alerts are particularly useful in critical operations that allow time for corrective action before some adverse consequence occurs. For example, a piping manifold may be designed such that a warning alarm sounds if valves are not opened/closed in the proper, critical sequence, or if a critical valve is left opened/closed.
- **Automatic System Shutdown.** Automatic system shutdown should be incorporated into the design when an illegal action is performed during a critical operation and no time is available for corrective action. For example, if a plant worker uses the wrong sequence in charging a reactor, the reactor will shut down before other materials are charged that may lead to critical temperatures and pressures.

F. FITNESS FOR DUTY

Successful task performance requires that the capabilities that plant workers bring to the task fall within an expected range. **Fitness for Duty** issues include reduction in a plant worker's mental or physical capabilities due to substance abuse, fatigue, illness, or stress, which increases the likelihood of errors. Types of possible impairments include:

- Physical attributes – strength, reach, eyesight and color acuity, hearing,
- Mental attributes – drug and alcohol (abuse), mental stress (on and off the job);
- Fatigue – issues from on the job and off the job (especially control of hours per work-day and per work-week)

Safeguards to prevent fitness for duty related errors include company programs for the detection and prevention of potential or actual impairment, as well as the individual responsibility of plant workers to decline assignments if they are impaired for any reason. The latter safeguard is a weak one, however, because humans are generally over-confident of their capabilities when under the influence of drugs or alcohol, or are stressed, fatigued or ill. Other factors that may discourage self-reporting include the fear of poor performance reports from supervision or not receiving pay for extra overtime.

Company programs that may be implicated in errors caused by plant worker impairment include:

- **Fitness for Duty Program** – Company fitness-for-duty programs are primarily responsible for detecting and preventing impaired plant workers from performing tasks that may affect public health and safety. Medical evaluations of personnel, behavioral observation programs, employee assistance programs and drug and alcohol testing are used to detect impairment. Impaired plant workers can be prevented from performing tasks by establishing protocols for instances in which a plant worker is believed to be unfit for duty, training for supervisors on detection of and response to fitness for duty issues, and employment guidelines for personnel who voluntarily enter drug or alcohol treatment programs. Weaknesses in a Fitness for Duty program may allow impaired personnel to have access to vital areas in a plant where they could commit errors. One excellent starting point for a Fitness for Duty sub-element is the guidance provided in ***US NRC's 10 CFR 26 (2005)***[15].

- **Overtime Policies and Practices** – Most companies establish limits for work hours to reduce on-the-job fatigue. It has been shown that 17 hours of work without a break results in the same error rate as being legally drunk. And, at 10 days straight of 12 hours work-days, the error rates for non-routine tasks such as startup of a continuous unit can increase to 1 mistake in 5 to 10 steps (as opposed to the target of 1 mistake in 100 steps). Routine authorization for work hours in excess of those recommended may result in fatigued plant workers. Further, a practice of excluding training or meetings that occur outside of a plant worker's normal work schedule from work-hour limitations will also contribute to fatigue. In ***US NRC's 10 CFR 26 (2005)***[15], the guidance given for control of overtime hours is no more 72 hours of work per 6-day period, no more than 16 hours of work in one day within that period, and a minimum of 24 hours contiguous hours away from work within a 7-day period. The US DOT has even more stringent rules on limiting work hours and establishing required hours for recovery from fatigue.
- **Shift Scheduling** – Shift scheduling may also affect the likelihood that plant workers will show performance decrements due to fatigue. A change in the assigned shift or a rotating shift schedule will disrupt circadian rhythms and may increase the likelihood of errors.
- **Safety Culture** – The effectiveness of self-reporting and behavioral observation programs depends greatly upon the safety culture at a site.

G. WORK PROCESSES AND SUPERVISION

Supervision is the process by which work is directed and overseen by first-line management. Successful supervision requires a combination of leadership skills and technical competence. Supervision differs from peer checking or quality control because a supervisor has line management responsibility for the plant worker(s) as well as responsibility for the work activity.

Supervision is more than the moment-to-moment direction of a work activity. Successful supervision requires the assessment and shaping of plant worker attitudes and motivation, communication and implementation of management expectations for performing work, the assignment of the best-qualified workers to various tasks, as well as the technical competence to identify incorrect actions and stop improper activities before an error is committed. Effective supervision involves directing the work, overseeing how it is performed, and leadership.

Organizations are typically structured to have sufficient supervision of the job, but many times the delineation between the trainer and the supervisor roles is blurred. Supervision can and normally does play a key role in selecting of the right worker for the job, scheduling of workers to match the required tasks for the day/week, and generally overseeing the task execution to ensure policies and procedures are followed. Supervisors are not always trained on all of their key roles in support of control of human factors, such as detecting issues in plant workers related to fitness for duty or fatigue.

H. WORK ENVIRONMENT

The **Work Environment** refers to the physical conditions in which work is performed. Environmental conditions that can affect performance include excessive vibration and noise, temperature extremes, and insufficient lighting, as summarized below:

- There are two types of **vibration** that may cause errors. The first is whole-body vibration, in which vibration is transferred to the plant worker from standing or sitting on a vibrating surface. The second is object vibration, in which a stationary worker interacts with a vibrating object in some fashion. The effects of vibration depend upon its frequency and acceleration. Frequency is the number of oscillations (cycles) that occur in one second. Acceleration is the force, or intensity, of the vibration.
- **Noise** is unwanted sound. Noise can cause errors in several ways. It may disrupt communications, affect the ability to perform tasks and annoy plant workers. The effects of noise on communications are complex. Even relatively low levels of noise can mask speech, but only under some circumstances. For example, speakers naturally raise their voices when there is background noise and may be able to overcome some of its effects on communication. Being able to see the speaker's face or using standardized phrases also improves communication in a noisy environment. The type of background noise also affects communication. It is easier to communicate over noise that is steady and uniform than noise that includes sharp tonal peaks, such as background speech.
- **Heat exposure** is a common problem in many areas of a plant, such as the turbine building when the plant is operating. The extent to which plant workers will be affected by heat depends on many factors. These include their physical characteristics, such as age, weight, acclimation to heat, physical fitness and dehydration. Other factors that determine the effects of heat on performance include airflow, humidity, clothing, and level of physical activity.
- **Exposure to cold** affects the performance of manual tasks. Decreases in the ability to control hand movements begin at an air temperature of approximately 54° F. The fingers may become numb to pain at this temperature and touch sensitivity is reduced. Performance of gross manual tasks, such as those involving the arms and legs is also degraded at 54° F. The speed at which manual tasks can be performed is affected by the rate of cooling. Slow temperature drops have a greater negative impact on manual dexterity than rapid temperature decreases, during the initial exposure period.
- **Adequate lighting** is required for accurate performance of nearly every task in a unit operation.

The organization must have engineering controls to help control each factor, but sometimes there is no other choice but to rely upon administrative controls.

I. COMMUNICATION (most importantly, verbal communication)

Communication is the exchange of information while preparing for or performing work. Verbal communication occurs face-to-face, by telephone, sound-powered phones, or walkie-talkies, as well as over public address systems. Written communication occurs, for example, through policies, standards, work packages, training materials, and e-mail.

Communication involves two sets of behaviors: (1) creating and sending messages and (2) receiving and interpreting them. Communication always involves at least two individuals, the sender and the receiver, and occurs:

- Between individuals
- Within and among work groups
- In meetings
- In pre-job or pre-evolution briefings
- During shift turnover

Successful communication requires several steps. The sender first develops the intention to communicate **either** verbally or in writing. The sender then composes a message that presents the meaning as clearly as possible. The receiver must pay attention to the message and then interpret its meaning. If the communication is successful, the receiver interprets the message consistently with the sender's intended meaning. However, there are many potential errors in both sending and receiving the communication:

- Sending Errors

    - *Content wrong*
    - *Content inconsistent*
    - *Content inappropriate for the job*
    - *Content inappropriate for the receiver*
    - *Standard terminology not used*
    - *Familiar terminology not used*
    - *Message production inadequate or interfered with*
    - *Necessary information not sent*
    - *Wrong place or person*
    - *Wrong time*
    - *Sending verification failure*

- Receiving Errors

    - *Information not sought*
    - *Information not found*
    - *Information not used*
    - *Receiving verification failure*
    - *Message misunderstood.*

Communication errors may be reduced in the following ways:

- Structuring messages in a standard format, which alerts the receiver if important information has been skipped
- Providing written instructions or procedures for complicated tasks
- Establishing protocols to repeat key parts of verbal communications
- Preceding special or nonroutine activities with a pre-job briefing
- Using structured protocols, checklists and logs to supplement written instructions

One key opportunity for communication errors is when a new shift takes over responsibilities from the previous shift. An effective tool to reduce these errors is a properly designed shift handover process that includes appropriate checklists and logs of activities, and a structured and consistent method for verbally reviewing this information. Additionally, the logs and discussions need to include:

- Routine operations
- Transition operations (e.g., changes to produce a different product or switch to a different raw material source)
- Nonroutine operations (e.g., maintenance activities or temporary operating conditions)

6. Establishing a Safety-Principled Organizational Culture

To ensure the success of the management systems created to improve Human Factors for plant workers, the proper culture of the organization must be developed so that this can be done effectively. Merriam Webster defines culture as "the set of shared attitudes, values, goals, and practices that characterizes an institution or organization". An organization's culture ultimately reflects the collective values of individuals and groups and can be influenced by the personalities of charismatic leaders, by shifting priorities based on the perceived scarcity of resources or actions necessary for survival, or by a sustained history of successes, behaviors, and beliefs. In an especially sound culture, deeply held values are reflected in the group's actions.

Organizational Culture is a somewhat intangible attribute of a company that is evident by the words and actions of management and all employees – including the plant workers. It is further defined informally as "how we do things around here", "what we expect here", or "how we behave when no one is watching". Reducing human error by plant workers therefore requires that the proper organizational culture be built and maintained.

A Safety-Principled Organizational Culture is one in which the actions of the entire organization are driven by the genuine desire to complete every task and achieve every objective in a manner that protects employees, plant equipment, and the community in which the plant is located to the greatest extent possible. While profitability and production targets are also important goals for the organization, the entire organization understands that these are enabled by continuous safe operations, rather than despite them.

Establishing a strong Safety-Principled Organizational Culture is a long-term process that requires dedicated resources for as long as the organization exists. Additionally, it is the result of the quality and quantity of effort put into its implementation. A solid, sensible effort by management and technical staff to establish a culture of this nature will be noticed by the plant workers and they will want to help. This will have the effect of catalyzing the program, thereby helping to ensure its success. Alternatively, merely stating the objective of building this culture will have little effect on creating excellent organizational safety culture if the implementation and control of safety is weak.

Establishing a strong Safety-Principled Organizational Culture will minimize safety incidents, and this will save lives. Since essentially all accidents begin with a human error, the actions or inactions of the plant worker can be the limiting factor in determining overall safety performance.

The proper culture helps plant workers understand why strict adherence to procedures is the right thing to do. Further, it helps to enable thoughtful compliance and to alert management when procedures are wrong. In this way, a sound culture is essential to maximizing the results of ALL safety activities.

Establishing a strong Safety-Principled Organizational Culture can be achieved by successfully pursuing 5 Key Strategies – each of which are described below.

Strategy 1: Establish Safety as a Core Value

Attaining the goal of superior plant worker performance (i.e., low human error rates) begins by providing strong and consistent communication about the importance of safety **from the very top of the organization**. This is achieved by setting an appropriate tone that safety is equal to or greater than production and by not diluting the message with a lot of other high priority core values. The first step in setting the tone is by clearly stating the importance of safety in a high-level mission statement. Then, this mission statement must be communicated to the organization, along with the related goals and plans for achieving it. An important aspect that must be communicated is the key role of controlling plant worker human error by focusing on Human Factors. Additionally, it is critical that this effort not be reduced to a "program of the month" – it must be spoken about frequently and consistently to all levels of the organization.

Of course, words are hollow without consistent corresponding actions. Therefore, strong leadership must be provided by demonstrating the commitment to safety excellence through actions and decisions. Replace slogans and banners with high standards, shared values, and core beliefs which sustain the organization regardless of who is in charge. To help ensure these values propagate throughout the entire organization, managers must be educated in this culture, vision, and related standards, as well as the expectations for their roles and responsibilities in this critical effort.

A strong Safety-Principled Organizational Structure relies upon high standards of performance that are consistently enforced. This is achieved by first defining appropriate safety goals – namely, those that provide an indication of the organization's capability to perform their work safely. Too often, organizations tie the success of their safety programs on the lagging indicator of incident rates. While this is of course the ultimate goal of the organization, this isn't necessarily a reliable indicator of safety for the organization. Consider how many ways a plant worker can be hurt completing a single task, then multiply that by the number of tasks the individual completes per day, and then by the number of plant workers within the organization. Clearly there are thousands of opportunities for injury each day. Fortunately, however, for nearly all organizations the number of incidents that occur is very small compared to the number of opportunities for an incident. Therefore, unless incident rates are abnormally high, a change in an organization's injury rate is typically due to factors outside of their control. It is much more valuable, then, to establish goals centered around the organization's work in improving their safety-related management systems.

Another aspect of establishing safety as a core value is to establish responsibilities and accountabilities for process safety roles. Every individual within the organization should be held accountable for steadfastly adhering to the newly established expectations. While it may be tempting to overlook seemingly minor deviations from these expectations, this can lead to normalizing deviations which can potentially lead to serious incidents due to a relaxed response to these warning signs of potential hazards. Therefore, the leaders of the organizations must hold direct reports accountable for demonstrating behavior consistent with this value and set expectations for each level of the organization to hold their direct reports to this same behavior.

Finally, it must be recognized that improving the organizational culture involve costs and benefits that are both long-term in nature and sometimes difficult to measure. By establishing appropriate goals as noted above that focus on management system health, however, the organization has the means to continuously improve its performance.

Completing all of the above will help ensure that the core value of safety is consistently embraced throughout the entire organization, including the plant workers.

Strategy 2: Provide the Necessary Resources to Ensure Success

Providing resources to transform the organizational culture is the most tangible way to demonstrate to the plant workers that safety is indeed an organization's core value. In this context, resources constitute monetary commitments, as well as those of personnel.

The place to begin is to fund and allocate personnel to enable good safety performance by creating and effectively implementing the Human Factor management systems described above. Done properly, these will close any gaps in the control of Human Factors as well as those evident in existing systems that don't meet a critical need such as performing a PHA of procedures to find scenarios unique to non-continuous modes of operation as described by Bridges and Marshall[26]. Adequate personnel must also be allocated for completing safety-related tasks to support these management systems (inspections, maintenance, etc.), for maintaining these management systems as changes or newly identified requirements occur, and for training and educating personnel.

It is also important to designate a high-ranking leader for safety who can help ensure that major decisions are based upon a proper consideration of safety. But it is also important to provide the positional support necessary to promote strong safety performance throughout the organization. This is achieved by identifying resources on paper and filling with effective and competent people. For example, if a trainer is shown on an organization chart, it should be ensured that this is their full-time job and that they aren't responsible for other unrelated tasks. This demonstrates that the organization places a high value on training and development of people and groups. This can be further enhanced by ensuring that training materials are maintained current, and by maximizing the use of instructor-led training, rather relying than too much on the use of Computer-Based Training.

Furthermore, a properly designed and implemented training program will allow the organization to demonstrate that it places a high value on safety knowledge and expertise. This can be enhanced by identifying and providing the proper mix and amount of expertise for its most important tasks (for example, by pairing an individual with a high level of expertise with one who has the potential, but not yet the experience, to develop this same level of expertise). To properly ensure optimum prevention of human error, the organization should focus on building expertise in Human Factors and seek out expertise for critical decisions.

Finally, since excessive overtime impacts employees' abilities to perform their jobs safely and increases safety risks, the organization must devote sufficient manpower resources to properly control overtime.

In summary, the overarching approach here is to ensure that a structured, strategic approach is developed to properly allocate funds and personnel to create, implement and maintain the management systems needed for safety enhancement opportunities.

Strategy 3: Establish a Blame-Free and Learning Environment

It is critically important that the organization commit to a blame-free culture for all plant worker human errors, including for near misses reported and investigation findings. At an intellectual level, everyone understands that no one is perfect and that zero human errors are not possible. Plant workers are no exception. A key driver for the creation of a blame-free environment then is driven by how the organization responds to the inevitable human errors that plant workers will make.

The response to plant worker errors can be thought of as a continuum that spans two extremes; on the left side the response is one of blaming or fault-finding, while on the right side the response is one of trying to understand or learn why the error was made. Quite often people will automatically default to the "blame" side of the continuum for plant worker human errors, saying something to the effect that if only the plant worker had handled the situation differently the error would not have occurred. While this is true at face value, it is not a helpful response. After the fact, it is easy to see what should have been done differently because the outcome is now known. But when the plant worker was taking actions, this outcome was not obvious to him or her. Significantly, people do not come to work planning to make errors; instead, they come to work to complete their tasks in the most efficient manner possible. It is only after an incident occurs that the benefit of alternate actions is clear. Thus, the more productive response is to try to understand the context of the plant worker's action – what was it that made this particular action appear to be correct? It is only by taking this approach that the lessons from the incident can be fully understood.

Each person in the organization needs to understand where on the human error response continuum they reside and make every effort to move as far to the right side as possible. As shown in Figure 5, there are three aspects to an individual's response to a plant worker human error: the **internal response** (what we say to ourselves when we are made aware of a human error), the **external response** (what

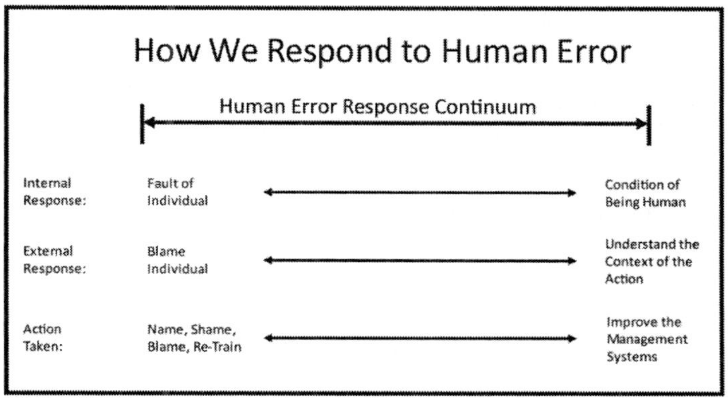

Figure 5: How We Respond to Human Error

we say out loud when we are made aware of a human error), and the **action taken** (what we do when we are made aware of a human error).

The internal response continuum ranges from thinking that the error was the fault of the plant worker, to recognizing that the error was a condition of that plant worker being human. The best approach for an individual to move to the more productive right side of this spectrum is to consider experiences in their own life in which they have made an error, which in retrospect seems to have been easily avoided. With this perspective, it is easier to understand that the plant worker who made the error in question was doing the best job possible but did not fully understand a specific hazard associated with it.

The external response continuum ranges from blaming the plant worker, to understanding the context of the action (i.e., why it made sense for the plant worker to take the action he or she took). Only by understanding the context of the action is it possible to develop measures so that another plant worker faced with the same choice does not repeat the error. As an example, a major chemical company experienced an acid spill from a tank truck one very cold evening. The truck driver was directed by the chemical company's guard to the wrong operating unit and began unloading the acid into that unit's tank. The operator from the unit soon emerged from the local control room and informed the driver that no acid was needed there. After a few phone calls, it was learned that the delivery was for a different unit which was located about ½ mile away. Accordingly, the driver disconnected his truck but failed to properly tighten the unloading hatch on top of the tank truck (most likely due to the cold ambient temperature). Therefore, as the truck was driven to the other unit, acid was spilled along the entire route. The chemical company's response to this incident was to inform the delivery company that that driver was no longer allowed on company property – in essence, blaming the driver for the spill. But this decision left many questions unanswered. Why was the driver directed to the wrong unit by the company's guard? Why was the driver allowed to unload directly into a company tank without checking in to the local control room (which is a requirement even for company employees visiting from a different plant area)? Only by answering these questions – understanding the context of the action – could assurance be provided that future spills of this nature would not occur.

The action taken response continuum ranges from a combined action of naming the plant worker responsible, shaming the person for making the error, blaming them for the consequences, and then re-training them, to making improvements to the underlying management systems. Using the acid spill example, the driver most certainly knows how to correctly complete all of the tasks needed to safely unload product; therefore, training would provide no real benefit other than to make him feel worse than he already did. However, by improving the management systems (strengthening the process for directing drivers to the proper locations, enforcing the rule to check in to the local control room before unloading) a better assurance that the human error could be prevented from recurring is possible.

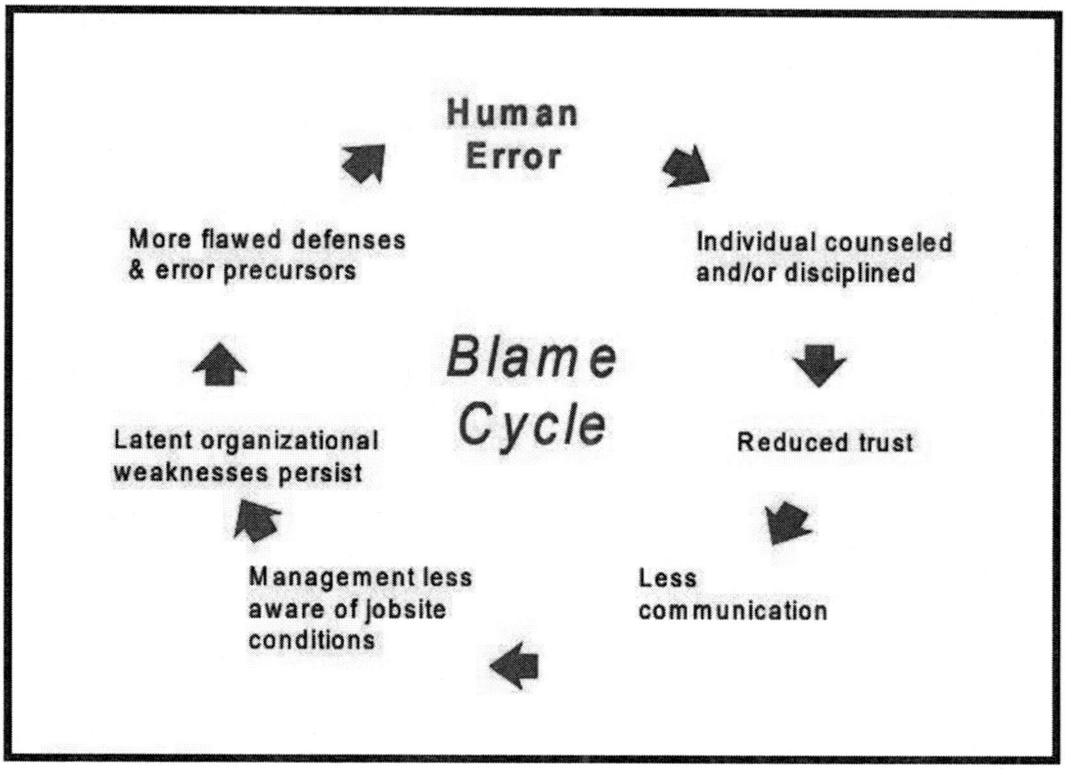

Figure 6: The Blame Cycle

The above discussion highlights the fact that residing on the "blame" end of the human error response continuum prevents an organization from truly learning about the root causes of incidents, thereby potentially leading to a recurrence of the error. However, blaming plant workers pose further risks to the organization. This is illustrated by the "Blame Cycle" that was developed by the US Department of Energy[27]. This cycle establishes a phenomenon that when a plant worker is blamed for a human error, this results in this individual ultimately not reporting future errors, which the organization is therefore unaware of and is unable to address. This then leads to future human errors due to these latent weaknesses, which are unreported, and the cycle repeats itself potentially leading to very serious consequences.

With this perspective then, a blame-free and learning environment can be achieved by ensuring that the first questions following an incident are what and why, not who. Those are followed by asking how we can improve the procedures, task design, practices, equipment, and other management systems to prevent this from happening again. With this environment, all incidents (especially near misses) are reported and investigated without blame, which results in root causes (management system failures) being identified, and corrective measures being taken to implement sustainable and permanent solutions. Trust and collaboration are high because the culture is blame-free when it comes to mistakes and people ask for help when they need it and give help without being asked.

Some key approaches to establishing a questioning/learning environment include properly performing risk reviews, conducting thorough incident investigations, performing table-top drills, and sharing and seeking lessons inside of and outside of facility. Significantly, plant workers should be highly involved in each of these activities. It is also critical to maintain a high awareness of hazards and consequences and remain vigilant for indications of system weaknesses. This can be achieved by ensuring that all plant workers are educated on the hazards of the operation, that they understand how safety hinges on control on Human Factors, that they all are educated on the consequences of deviation from safe practices, and that they share lessons learned.

A final aspect is to provide timely response to, and communication of, safety issues and concerns and lessons learned from incident investigations, audits, risk assessments, employee concerns and discrepancies between practices and procedures.

In summary, an organization must ensure that a blame-free and learning environment is consistently embraced throughout the entire organization. This can be achieved by reinforcing this at all opportunities through words and actions by personally seeking solutions instead of blame. Furthermore, top management must hold direct reports accountable for demonstrating behavior consistent with this value and set expectations for each level of the organization to hold their direct reports to this same behavior.

Strategy 4: Enhance Employee Empowerment and Involvement in Safety

Plant workers can best contribute to the Safety-Principled Organizational Culture and thus successfully fulfill their safety responsibilities by having the necessary authority and accountability for safety responsibilities. Leadership can enable this by reinforcing that all employees have responsibilities to themselves, fellow employees, company, and community and by involving the plant workers in safety management systems to ensure that they are widely understood and supported. This approach has the added benefit of recruiting additional resources to support the many tasks required.

A prerequisite to this is for the organization to ensure open and effective communication – both vertically and horizontally. Strategically, this can be achieved by emphasizing the need to promptly observe and report non-standard conditions. A key management system to enable this is a properly deployed near miss reporting program. Bridges[28] outlines an approach for creating a program of this nature, as well as specific, common hurdles that must be overcome to ensure its success. Key items in a properly designed program include first providing means for all plant workers to communicate hazards to management (in a blame-free environment) and for management to consistently act on these stated concerns.

Related to this, plant workers should be involved in identifying hazards and deciding how they should be addressed. In certain situations, plant workers should be authorized to take action to address hazards without management involvement. A very important example of this is that plant workers should have the authority to question decisions which are being made about process safety, including a "Stop Work Authority".

Ensuring optimum plant worker involvement in the safety management systems must be intentional. Therefore, the organization should develop an Employee Participation Plan, which lists key components of the management system that plant workers can best contribute to. A few examples to include identifying and controlling deficiencies in Human Factors, writing and reviewing operating and maintenance procedures, and participating in Process Hazard Analyses, Root Cause Analyses, and audits. Other opportunities can be developed by critically reviewing components of all safety management systems (including those developed for Human Factors) that will benefit by having the specialized knowledge of plant workers.

Strategy 5: Drive Continuous Improvement in the Safety Management Systems

Even the best designed safety management systems are not perfect (because they are developed and managed by humans). Therefore, the best Safety-Principled Organizational Cultures undertake specific activities to understand gaps in the systems and then take actions to close these gaps. This is achieved by establishing relevant metrics on the health of the management systems, which as noted previously, should predominately be leading metrics. Typical examples include the timeliness of correcting deficiencies, adherence to ITPM schedules/procedures, and the expansion of problems detected in one area to investigate similar applications in other areas.

Over time, trends should be developed of these metrics, presented to management who should then direct actions to improve these Key Performance Indicators. Further, these metrics should be shared within the organization so that all employees – including plant workers – have a clear sense of how well their efforts to contribute to the success of these programs are working. It is important to remember that many of these safety-related metrics have long-term results, rather than an immediate impact. Therefore, continued diligence in these continuous improvement efforts are needed to ensure a sustained Safety-Principled Organization Culture, and ultimately outstanding safety performance for the organization.

7. Closing

Human error is inevitable. And the human errors that are typically thought of as the ones that lead to serious accidents are often attributable to those made by plant workers. Although this assessment is unfair (as other members of the entire organization contribute in meaningful ways to these incidents), it is nonetheless very important for organizations to undertake measures to reduce plant worker human errors.

The means to do this are outlined in this paper. It relies first on the organization having the proper culture – one that doesn't react to plant worker errors by blaming them; rather, it seeks to understand why the error was made and taking actions to fix the issue instead of the plant worker. It recognizes that to do this requires that the error must be traced to the underlying Human Factors that increased the likelihood of this error's occurrence, and that the best way to improve these Human Factors (and thus reduce plant worker human errors) is to develop and implement robust management systems for them.

The information provided in this paper to strengthen the plant worker Human Factor management systems is the most important method to reduce plant worker human errors. However, even with perfectly implemented management systems human errors will still occur – albeit at a lower frequency and likely with a smaller adverse impact than before. Therefore, in cases in which the consequence of a human error is unacceptably high, the organization should implement solutions that will lower the residual risk further. These solutions can be designed to either help prevent the error or compensate for the error, but **they must be independent of the human error initiating event**. Bridges and Bridges [25] have compiled a list of the best remedies for human error for these situations. In a similar vein, plant worker human error can be further reduced by performing the previously mentioned PHA of procedures to find scenarios unique to non-continuous modes of operation as described by Bridges and Marshall[26].

Finally, a small fraction (about 15%) of plant worker human error is attributable to acquired habits, or behaviors; the remedy for which is a properly designed and implemented plant worker observation program. The reader is encouraged to review these other important documents and methodologies to obtain a fully comprehensive understanding of the best ways to reduce plant worker human errors.

8. References:

1. Swain, A. D., Guttmann, H. E., *The Human Reliability Handbook - with Emphasis on Nuclear Power Plant Applications*; NUREG/CR-1278, U.S. Nuclear Regulatory Commission, Division of Facility Operations, 1983.

2. Gertman, D.; Blackman, H.; Marble, J.; Byers, J. and Smith, C.; *The SPAR-H Human Reliability Analysis Method,* NUREG/CR-6883, U.S. Nuclear Regulatory Commission, Office of Nuclear Regulatory Research, Washington, DC, August 2005.

3. Bridges, W., and Tew, R., *Human Factors Elements Missing from Process Safety Management (PSM),* 6th GCPS, AIChE, March 2010.

4. *Risk-Based Process Safety*, CCPS/AIChE, John Wiley & Sons, Inc.; Hoboken, NJ, 2007.

5. *Conduct of Operations and Operational Discipline*, CCPS/AIChE, John Wiley & Sons, Inc.; Hoboken, NJ, 2011.

6. *US FAA Advisory Circular* 120-42A (as part of ETOPS approval)

7. Bridges, W. G., Williams, T. R., "Effective Procedure Writing – A Two-Part Series (Part 1: Reducing Risk of Worker Errors; Part 2: Addressing Process Safety Management Requirements for Operating Procedures)," *Chemical Engineering Progress*, 1997.

8. Bridges, W., and Collazo-Ramos, G., *Human Factors and their Optimization,* 8th GCPS, AIChE, April 2012.

9. *Guidelines for Hazard Evaluation Procedures*, 3rd Edition, CCPS/AIChE, John Wiley & Sons, Inc.; Hoboken, NJ, 2008.

10. Bridges, W., *Optimizing Hazard Evaluations,* LPS/AIChE, April 2009.

11. Bridges, W., Lorenzo, D., and Kirkman, J., *Addressing Human Errors During Process Hazard Analyses,* CCPS Conference, AIChE, 1991 (also published in Chemical Engineering Progress May 1994).

12. *Guidelines for Investigating Chemical Process Incidents, 2nd Edition,* CCPS/AICHE, John Wiley & Sons, Inc.; Hoboken, NJ, 2003.

13. Bridges, W., *Gains in Getting Near Misses Reported,* 8th Conference, ASSE-MEC, Bahrain, 2008.

14. Della Rocco, P.S., et al., *The Effects of Napping on Night Shift Performance,* Federal Aviation Administration, U.S. Department of Transportation, 2000.

15. *Fitness for Duty Programs, Preamble, 10 CFR 26, U.S. Nuclear Regulatory Commission, Revised 2005.*

16. US Chemical Safety and Hazard Investigation Board (CSB), *Anatomy of a Disaster: Explosion at BP Texas City Refinery,* published 2006.

17. University of South Australia (UniSA), Adelaide Centre for Sleep Research, *SDA Fact Sheet - AT03*, 2006.

18. Hallbert, B.; Whaley, A.; Boring, R.; MCabe, P.; and Chang, Y.; *Human Event Repository and Analysis (HERA)*, NUREG/CR-6903, U.S. Nuclear Regulatory Commission, Office of Nuclear Regulatory Research, Washington, DC, 2007.

19. Bridges, W., and Clark, T., *LOPA and Human Reliability - Human Errors and Human IPLs,* 7th GCPS, AIChE, March 2011.

20. Bridges, W., and Thomas, H., *Accounting for Human Error Probability in SIL Verification Calculations,* 8th GCPS, AIChE, April 2012.

21. Bridges, W., *The Impact of Human Factors on LOPA,* 9th GCPS, AIChE, April 2013.

22. Bridges, W., and Tew, R., *Best Practices for Writing Operating Procedures and Trouble-shooting Guides,* 13th GCPS, AIChE, March 2017.

23. Bridges, W., *Proven Approaches to Ensuring Operators Can Respond to Critical Process Deviations in Time (Human Response IPL),* 13th GCPS, AIChE, March 2017.

24. Bridges, W. and Rhodes, W., *Everything you need to Know about Human Reliability and Process Safety,* 17th GCPS, AIChE, March 2019.

25. Bridges, S. and Bridges, W., *Abnormal Modes of Operation – A risk-based focus,* 16th GCPS, AIChE, March 2020

26. Bridges, W. and Marshall, M., *Necessity of Performing Hazard Evaluations (PHAs) of Non-normal Modes of Operation (Startup, Shutdown, & Online Maintenance)*, 12th GCPS, AIChE, April 2016

27. DOE Standard – *Human Performance Improvement Handbook, Volume 1: Concepts and Principles*

28. Bridges, W., *Gains from Getting Near Misses Reported*, 8th GCPS, AIChE, April 2012

619491 Lessons-Learned from Remote HAZOP/LOPA Applications in 2020

Abstract:

Going into Year 2020, basic brainstorming team session approaches for the performance of a Hazard and Operability (HAZOP) Study or a Layer of Protection Analysis (LOPA) had not changed significantly over the years; however, the infection prevention measures necessary to address COVID-19 resulted in a rapid transition to widespread application of telecollaboration tools that previously had only been applied sporadically. Prior to 2020, telecollaboration tools for Process Hazard Analysis (PHA) had been used to short-circuit the need for pulling together international teams, for very short PHAs or MOCs, and other select applications. As 2020 drew to a close, numerous applications of HAZOP/LOPA Team Exercises using telecollaboration tools resulted in valuable insights into tools and techniques that were productive for different PHA applications. Properly applied, applications ranging from shorter Process Hazard Analyses (PHAs) and Management-of-Change (MOC) evaluations to longer studies addressing Five-Year PHA Updates and Capital Projects can be implemented productively using telecollaboration approaches.

Post COVID-19, many PHA applications will still be best addressed with the HAZOP/LOPA Team being co-located, but the savings in travel costs, the logistics associated with personnel travel, and the ability to quickly convene and implement a HAZOP/LOPA review can allow remote HAZOP/LOPA, through the use of contemporary technologies, to be an attractive option that can increase overall efficiency, without compromising quality. Experiences in the successful application telecollaboration tools in 2020 mean that Remote PHAs are here to stay.

A core objective of this paper is to provide insights into good applications and tips that can directly be used by the HAZOP/LOPA Facilitator to optimize the logistics to get the most out of a HAZOP/LOPA, where some or all team members are geographically distant.

SPRING21 +17TH GCPS
A Joint AIChE and CCPS Meeting

Research on Human Resource Development Exercises for Resilience to Cyber Incidents in Critical Infrastructure

Yuitaka Ota
Nagoya Institute of Technology,
System Management and Engineering
Showaku, Gokiso, 4668555 Aichi, Japan
cjr17009@nitech.jp

Erika Mizuno, Asai Haruna, Tomomi Aoyama, Yoshihiro Hashimoto, Ichiro Koshijima
Nagoya Institute of Technology,
System Management and Engineering
Showaku, Gokiso, 4668555 Aichi, Japan
ckv17032@nitech.jp, h.asai.852@nitech.jp, aoyama.tomomi@nitech.ac.jp,
hashimoto@nitech.ac.jp, koshijima.ichiro@nitech.ac.jp

Prepared for Presentation at
American Institute of Chemical Engineers
2020 Spring Meeting and 17th Global Congress on Process Safety
Virtual
April 18 - 22, 2021

Research on Human Resource Development Exercises for Resilience to Cyber Incidents in Critical Infrastructure

Yuitaka Ota
Nagoya Institute of Technology,
System Management and Engineering
Showaku, Gokiso, 4668555 Aichi, Japan
cjr17009@nitech.jp

Erika Mizuno, Haruna Asai, Tomomi Aoyama, Yoshihiro Hashimoto, Ichiro Koshijima
Nagoya Institute of Technology,

Keywords: OT-Security, cyber incident exercise, business continuity management

Abstract

Current cyber-attacks target IT systems and OT (Operational Technology) systems that are indispensable for the automatic operation of critical infrastructure industries. There have been various reports of cyber incidents targeting the OT system. In 2019, Norsk Hydor was infected with particular ransomware worldwide through its corporate network, forcing the manufacturing site to operate manually. In May 2020, Taiwan's CPU, responsible for Taiwan's oil supply, was targeted by a cyber-attack.

COTS (Commercial Off The Shelf, such as Windows OS, Intel PC, and open source applications) devices have been proposed to reduce the deploying OT system budget. Also, DX efforts are being made to shift from physical operation to virtual operation by using virtualization with IoT, AI, and Cloud. Due to this technological trend, the expanded attack surface is forced to manage the OT system as a "Zero Trust System" that assumes cyber-attacks. In IT systems, countermeasures such as whitelisting and firewalls prevent attackers from outside the system. For OT system, Network Security Zoning described in IEC62443 is recommended to mitigate the effects of cyber-attacks.

However, attackers exist all over the world, and they are continually developing new attack methods. For this reason, the authors believe that it is impossible to prevent all attacks, no matter what measures the defenders take. Therefore, when responding to cyber-attacks against the OT system, it is essential to consider countermeasures by focusing on plant accidents caused by control system failures, most of which are assumed as safety measures.

To ensure the plant's safety and continue the business operation even if it has a cyber-attack on the OT system, it is indispensable to develop human resources to respond to the incident flexibly. For this reason, we had been developing cybersecurity exercises to acquire the meta-knowledge necessary to respond to cyber-attacks targeting the OT system.

In this paper, the authors present the design framework, the implementation method, and OT expert's evaluation results of a proposed exercise.

1 Introduction

1.1 Background

An industrial control system (ICS) is a device or group of devices for managing and controlling other devices and systems. ICS is widely used in the Critical Infrastructure (such as electric power, city gas, and transportation) and the manufacturing industry that supports our daily lives, from those that are familiar to us. The conventional ICS has an isolated network as an independent system built with its hardware, software, communication protocol and has been considered resilient to cyber-attacks. However, in recent ICS, the adoption of COTS (Commercial Off The Shelf, such as Windows OS, Intel PC, and open source applications) devices is progressing due to versatility and budget. Also, DX (Digital Transformation) is being promoted for the purpose of plant operation support and operation optimization using IoT (Internet of Things) and AI (Artificial Intelligence) through open networks with the outside. As a result, ICS is exposed to the threat of cyber-attacks, and companies using ICS need to recognize not only the risk that has been assumed as the risk of plant shutdown but also cyber-attacks as one of the risks.

Cyber-attacks targeting ICS have been increasing in recent years. In 2010, malware called Stuxnet attacked Iran's uranium enrichment centrifuge [1]. Hackers caused power outages in Ukraine in 2015[2] and 2016[3]. In 2019, Norsk Hydros in Norway, the world's leading manufacturer of rolled aluminum products, was affected by cyber-attacks in plant operations [4]. In addition, the Information Security Agency (BSI) of the German Federal Government has published "Industrial Control System Security: Top 10 Threats and Countermeasures" [5].

In this way, cyber-attacks are a risk that threatens the safety of the plant, and companies must respond as a risk that threatens their business.

1.2 Response to a cyber incident

1.2.1 Cyber-attacks targeting Industrial Control System

Fig. 1 shows a cybersecurity testbed and a virtual ICS network. The testbed (Fig. 1) heats water with a heater in the lower tank and circulates the heated water through a simple pipeline to the upper tank (using a pump). The testbed is also equipped with actual industrial control devices and is controlled automatically[6]. In this ICS network, OLE for Process Control (OPC) Servers collect and exchange process data and monitor them by using Supervisory Control And Data Acquisition (SCADA) function included in the OPC Servers. OPC Servers receive various parameters from the testbed's sensor, and SCADA gives instructions to the control equipment using the parameters' data.

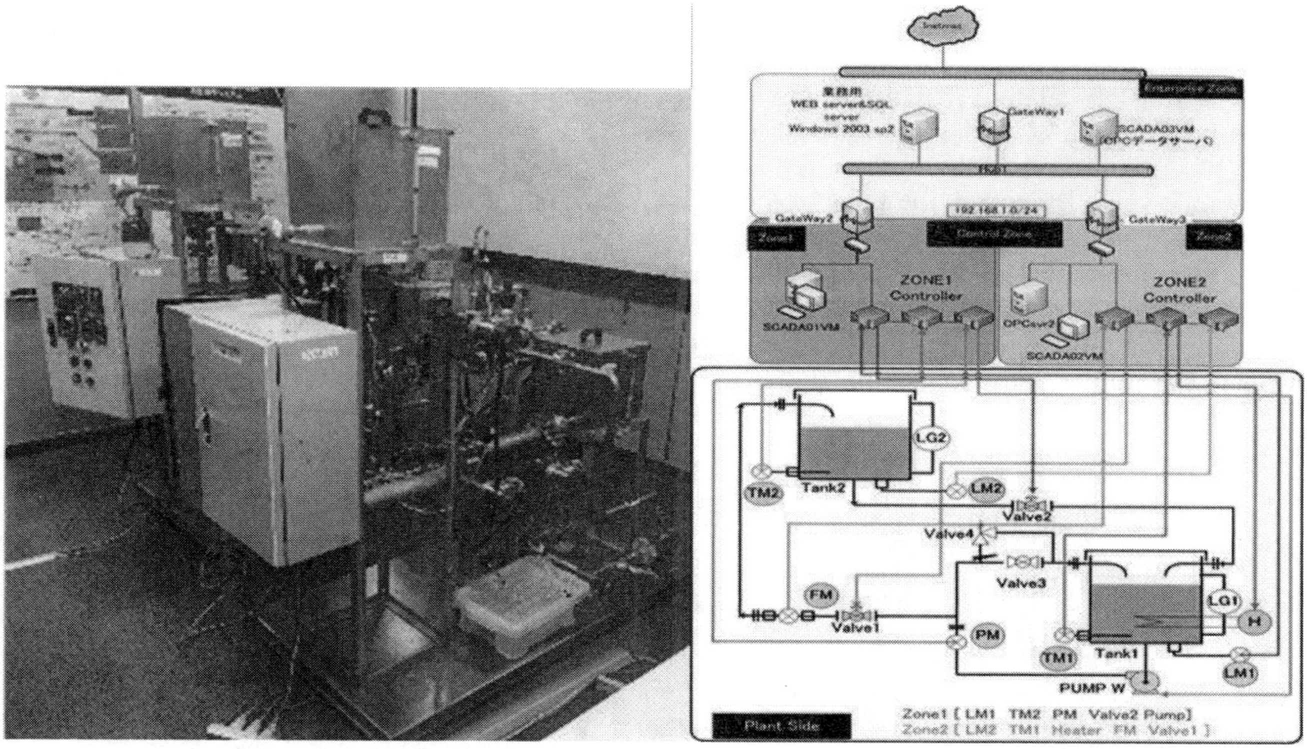

Figure 1. Cyber Security Testbed and ICS network

The authors have demonstrated cyber-attacks against the testbed, according to the steps of the anatomy of a cyber-attack. The attacking procedure is determined as follows.

1) Attack Step 1: Reconnoitering and enumeration

The attacker executes the Armitage to launch the Metasploit framework. Then, the attacker executes the Nmap, which scans a target network (ICS-2 Network 192.168.12.0/24). After completing the scan, the attacker gets target information.

2) Attack Step 2: Intrusion and advanced attacks

According to the functions of the Armitage, the attacker ascertains an attack method and vulnerability in the Metasploit framework. The attacker exploits payloads, which execute inappropriate code in the use of the vulnerability. After sending the payloads, the attacker obtains the administrator's permission for the OPC2 illegally.

3) Attack Step 3: Malware insertion

The attacker uploads a malware execution file to an OPC2 background process to execute a malware file. The configuration file of OPC2 was rewritten, and then the OPC2 was restarted. They were resulting from the attack, the valve controlled by the OPC2 shutdown.

4) Attack Step 4: Clean-up

The attacker deletes the malware file and the configuration file from OPC2 storage to not leave the cyber-attacks evidence and finish the cyber-attack.

As the valve is opened, the stage of the testbed changes. Usually, the testbed's state transition appears on the monitor of the operator (Fig. 2). The changing graph shows the liquid level of the upper tank. The liquid level is changing due to the malfunction of the valve. The operator can notice the testbed abnormality due to the unusual process deviations from the steady-state condition through the monitor. However, the attacker conceals the information about cyber-attack in order not to let anyone notice the cyber-attack. The testbed monitoring screen is concealed, as shown in Fig. 3, while the cyber-attack causes severe accidents at the testbed.

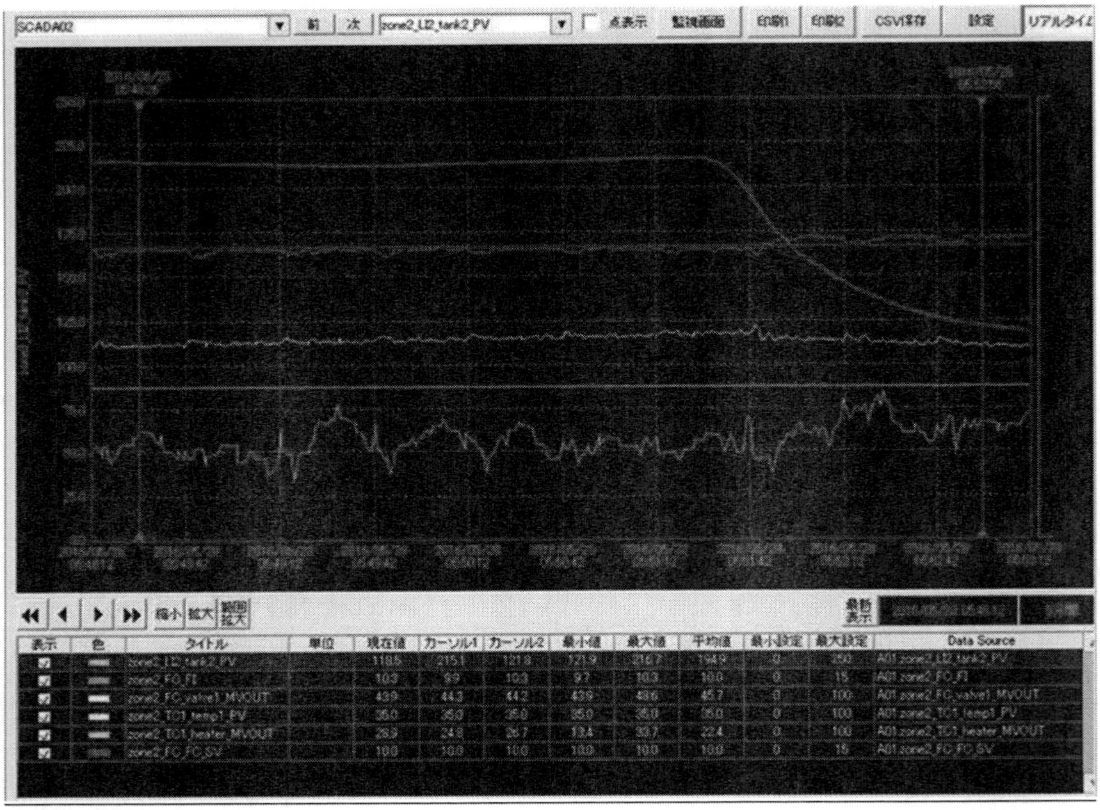

Figure 2. HMI Screen Shot in Operation (Safety Incident)

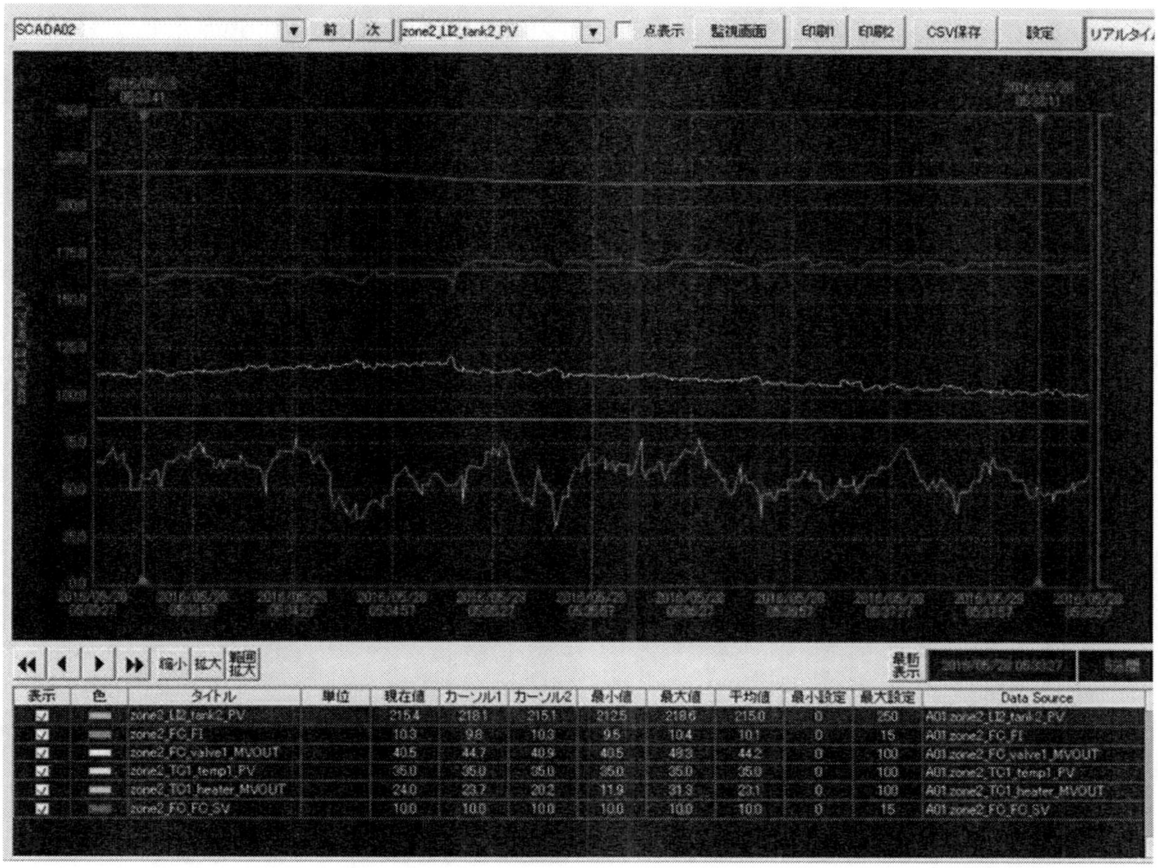

Figure 3. HMI Screen Shot in Operation (concealed by the attacker)

This demonstration shows a possibility of safety incidents caused by cyber-attacks to ICS.

1.2.2 Response to Cyber Incidents

The consequences of a cyber-attack depend on the characteristics of the plant. For example, an attacker hijacks a data server (DCS, OLE for Process Control server, etc.) that gives parameters and commands to control devices such as a PLC (Programmable Logic Controller) and gives the control devices malicious instructions. Such an attack can lead to abnormal operation and an incident, similar to a data server or control equipment failure or operator error with traditional equipment. That is, incidents caused by cyber-attacks in ICS systems are incidents that have already been considered in HAZOP scenarios with different initiating events. Operating companies have taken countermeasures to ensure the safety of the plant for many years. Companies are preparing countermeasures to ensure safety (Safety Response) under an incident at their plant. Human resource development is also carried out based on the countermeasures to respond when an incident occurs. In other words, even if an abnormality occurs due to a cyber-attack, it is possible to ensure the safety of the plant if the operation staff takes a proper Safety Response. However, problems peculiar to cyber-attacks, such as rewriting server configuration files and rewriting control device programs, may hinder the implemented safety-based Safety Response. Figure 4 shows the incident response structure that takes into account the cyber-attacks proposed here. If the plant is operating abnormally caused by the cause considered so far, it is possible to restore plant safety by performing a safety response, as shown in the left figure. However, there is a possibility that cyber-attacks will interfere with safety response (e.g., concealment of a plant's information on the monitor). Companies must consider where cyber-attacks interfere with safety response. Moreover, it is necessary to take the responsibility to eliminate the effect of cyber-attacks (security response) to take safety response as shown in the right figure. Therefore, if a cyber-attack causes a plant's abnormal operation, a response to rectify the situation is required in combination with a structure to add a security response to the safety response.

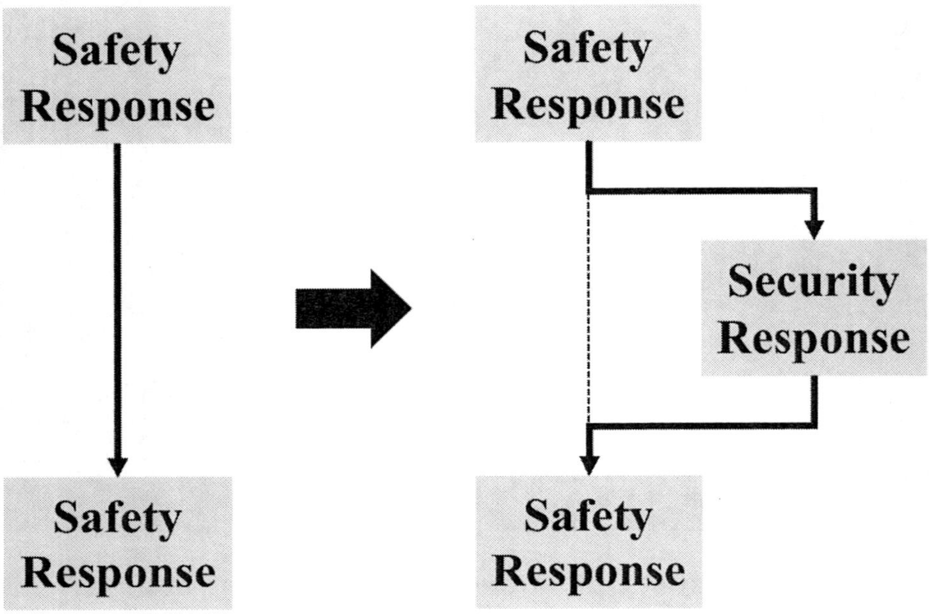

Figure 4. Incident Response Structure[7]

Responding to cyber-attacks targeting ICS systems, it is necessary to integrate a Safety Response to ensure the plant's safety and a Security Response to eliminate the effects of cyber-attacks. It is essential to create a plan that considers security incidents in addition to the existing safety response plan, and make it known to each operations staff.

1.3 Required organizational cooperation

The right staff needs to take action at the right time. To respond to incidents competently, it is necessary to properly share the current situation and communicate it among the right staff.

When an incident occurs due to cyber-attacks at a plant, two responses, a Safety Response to ensure the plant's safety and a Security Response to remove the cyber-attack effects, proceed simultaneously. Since the Safety Response responds to ensure the plant's safety, the response's responsibility is on the OT side (operators working at the plant). Since the Security Response is a response to remove the effects of cyber-attacks, the response's responsibility is on the IT side (IT engineer). Also, considering a cyber-attack on ICS, the result of the cyber-attack appears in the plant, and the cause is the IT network. In other words, it is a structure in which the cause-effects occur in different places. Also, in the incidents that we have experienced so far, there was no problem in dealing with the cause, but in cyber incidents, the same incident may occur in all parts connected to the Internet. Therefore, information sharing is vital for appropriately responding to the above-mentioned new type of incident, "Simultaneous Multiple Accidents."

2 How to develop Security Personnel

2.1 Preparation required as a company

・Formulation of ICS-BCP

Many companies have created a business continuity plan (BCP) in advance so that their business can continue even during natural disasters such as earthquakes and typhoons. It is important to develop a similar program for cyber incidents. It is necessary to analyze the risks in the event of a cyber incident and formulate an ICS-BCP that includes countermeasures for reducing, holding, avoiding, and mitigating the effects of the risks.

・Human Resource development

Even in unprecedented situations such as cyber incidents, companies must respond to continue their business. And it is the employees who respond to the incident. If the personnel's ability to respond to the incident is not sufficient, the company's damage will increase.
In other words, as a company, it is required to develop human resources who will respond.

2.2 Importance of virtual experience

As mentioned in 2.1, companies must prepare a BCP in advance for cyber incidents and develop human resources to carry it out. However, many companies have no cyber-attacks experience with ICS, and it is not easy to imagine how cyber-attacks may affect companies.

A mechanism is needed to experience cyber incidents and discuss necessary countermeasures.

As an approach to that mechanism, the authors chose an exercise. Figure 5 shows the flow for developing a training exercise. By analyzing incidents in the real world, we can extract the causes of incidents and the commonality of cyber-attacks targeting ICS. By giving the extracted information rules and game characteristics, the exercise participants will learn systematically. Thus, participants experience cyber-attacks on ICS through exercises.

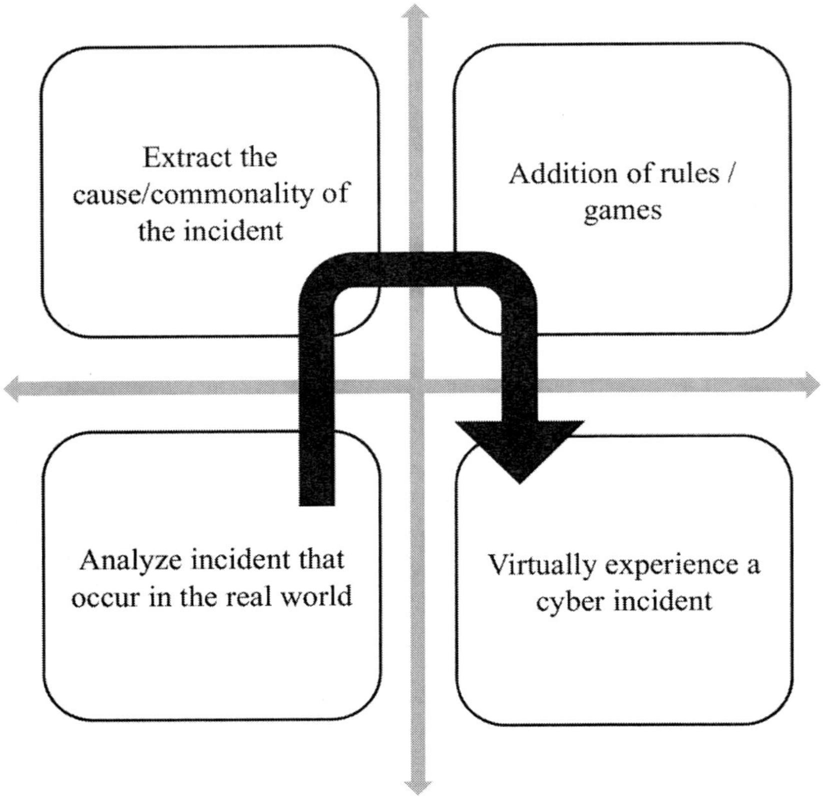

Figure 5. Exercise development process

Even staff who have never experienced a cyber incident can increase their experience with cyber incidents through exercises. By expanding this experience value, staff will more likely think and act even if people have a similar incident or an incident that people have never experienced before.

2.2 Human resource development method using virtual experience (exercise)

By taking the following four steps, we aim to improve the incident response capability of the company.

1. Information input
   Many companies have no experience in responding to cyber incidents, so it is necessary to understand the impact of cyber incidents on companies and plants. It is also required for the parties concerned to understand that cyber-attacks on plants are problems on the IT side and OT side.

2. Create ICS-BCP from a management perspective
   Companies are required to generate stable profits. When an incident occurs, companies need to think about what to prioritize to keep the business running, when to recover, and alternatives if there is no recovery. Each company must create an ICS-BCP assuming a cyber incident and communicate it to employees.

3. ICS-BCP with added on-site perspective
   It is necessary to respond to incidents based on the ICS-BCP defined as a company created in step 2. However, this BCP only includes the perspective of how to respond as an organization.

   By checking the BCP from the viewpoint of the operational site, it can be made into an ICS-BCP that can be operated.

4. Share the idea of all layers.
   Employees must understand ICS-BCP to respond even if a never-experienced-before incident occurs. Therefore, by conducting training centered on the planned ICS-BCP, we will deepen the understanding of employees and objectively evaluate the ICS-BCP.
   By repeating this training multiple times, it is possible to confirm omissions and deficiencies in BCP and increase ICS-BCP accuracy.

The activities mentioned above don't do only once. Each company can improve its resilience ability by making assumptions based on various preconditions (scenarios), rotating the PDCA cycle, and continuing discussions.

3 Exercises for improving resilience developed

Based on the recognition that not only hard skills but also soft skills are important for responding to cyber incidents, we have developed discussion-based ICS security exercises.

3.1 Theme of the exercise

As mentioned above, it is very important to prepare a response plan to respond to an incident with hard and soft skills. Therefore, the purpose of the exercise is to understand the required actions to be taken by cyber-attacks and the importance of inter-organizational collaboration needed to take those actions.
The final goal is to be able to create an ICS-BCP using the knowledge gained from the exercises.

In order to make participants understand the purpose of the exercise, the exercise was divided into three phases according to the purpose of correspondence. (Fig.6)
When an incident occurs in a plant, whether it is a cyber incident or not, it is necessary to take measures to ensure the safety of the plant. Then, it is necessary to investigate the cause after ensuring the safety of the plant. If the cause is a cyber-attack, it is necessary to investigate the extent to which the impact of the cyber-attack has spread.
Initially, the safety workload is large, but the security workload gradually increases.
Then, there will be a response to confirm the extent of the impact of cyber-attacks not only on plant safety but also on business continuity.

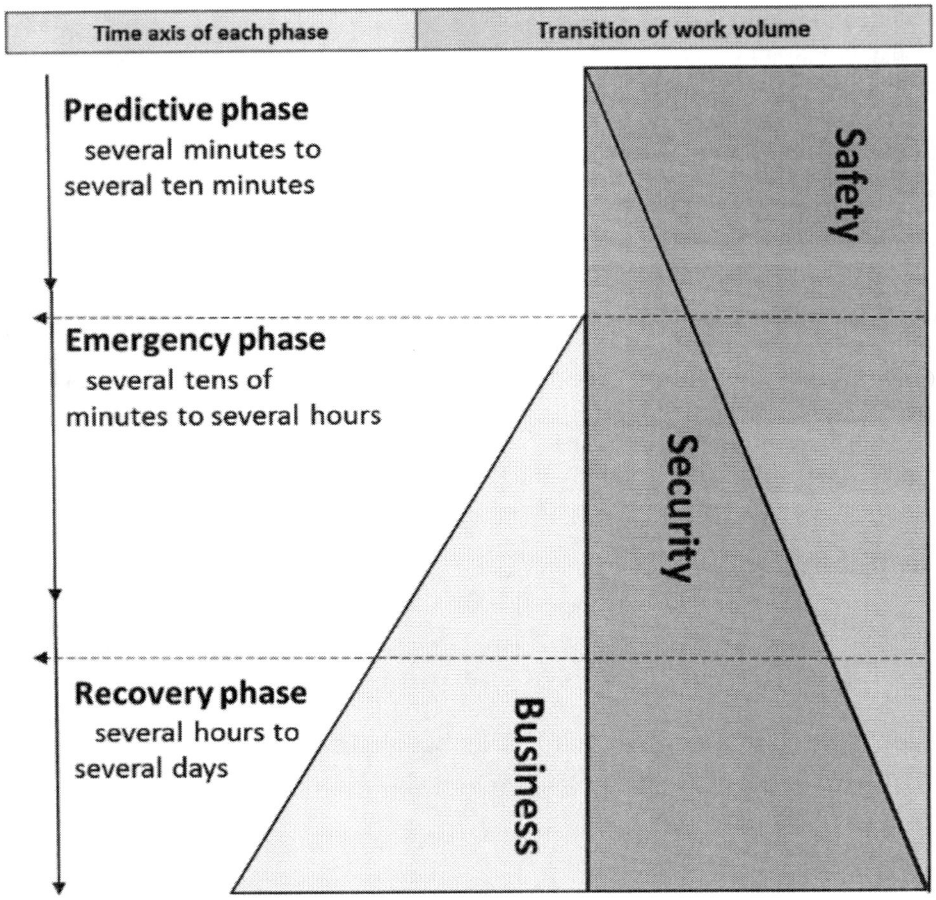

Figure 6. Structure of Exercise with Workload Image among Safety, Security, and Business Continuity Activities[7]

The authors visualized the state that this working volume change with the progress of time. (Fig6) Fig.6 expresses the work volume at a certain spot. The work volume is divided into safety, security, and business. The participant understands a change of this working volume through exercise.

1) Predictive Phase
The purpose of this phase is to ensure the safety of plants threatened by cyber-attacks. In this phase, the participants examine the safety response obstructed by a cyber-attack.
After detecting suspicious communication in a section that monitors the network, the exercise begins with an alert for the status of the liquid in the plant. In this phase, the participants must ensure the safety of the plant. Once safety is ensured, participants must investigate the abnormal operation. In this phase, participants can recognize how different causes can influence the safety response.

2) Emergency Phase
The purpose of this phase is to continue the business. The participants consider the security of the plant and how the exercise implicates a real cyber-attack. They must also take into account

how their business will be affected by the cyber-attack. In this phase, the participants consider practical measures taken to ensure business continuity throughout the company.
This phase begins after the abnormal operation of the plant reveals a cyber-attack. In this phase, participants can recognize that the entire company must be mobilized to treat a situation caused by a cyber-attack.

3) Recovery phase
The purpose of this phase is to restore own business to normal operation rapidly. The participants must find a way to nullify the influence of the cyber-attack so that the plant can return to normal operation. In addition, they must consider how business is affected by the temporary plant shut down. In this phase, the participants examine the start-up procedure of a plant crippled by a cyber-attack. In addition, the participants examine methods to prevent the recurrence of similar cyber-attacks.
This phase begins after the plant is stopped by a cyber-attack. In this phase, participants think about how to restart the plant. In this phase, participants can see how different causes affect the restoration of normal plant operations.

The following three types of training exercises have been developed.

- Card-Type Incident Response Exercise
- BCP Creation Exercise
- Simulation Exercise called "IMANE" (Incident MANagement Exercise)

3.2 Core scenarios for exercise

The developed exercises are based on the same scenario with different methods.
The base scenario is required to be able to consider the ideal organizational cooperation and necessary soft/hard skills for the organization to respond quickly in the event of a cyber-attack through exercises. The created scenario was composed of three scenarios to think about the three aspects of safety, security, and business.

Virtual Company:
Participants from various companies, departments have different backgrounds. A common scenario is indispensable for the participants to share the exercise together. In addition, using a simple and realistic virtual company makes it easier to apply the awareness of the exercise to the business. Figure 7 is an organizational chart of the created virtual company. Practical prerequisites, such as the company's business profile and its organization's intercommunication rules, have been set to achieve the exercise's objectives. Participants can discuss how information should be shared and how an organization should respond to cyber-attacks. Figure 8 is the network structure of a virtual enterprise. The attacker will attack using the computer that was hijacked by the cyber-attack as a stepping stone. The attacker steps on the hijacked computer and moves to the computer he wants to attack. In other words, the effects of cyber-attacks are spreading not only to the attack target but also to the attack process. For mitigating the impacts of cyber-attacks, it is necessary to consider countermeasures within the range of the effects of cyber-attacks. By conducting exercises based on this network structure, participants can discuss the Security Response required to respond to cyber incidents.

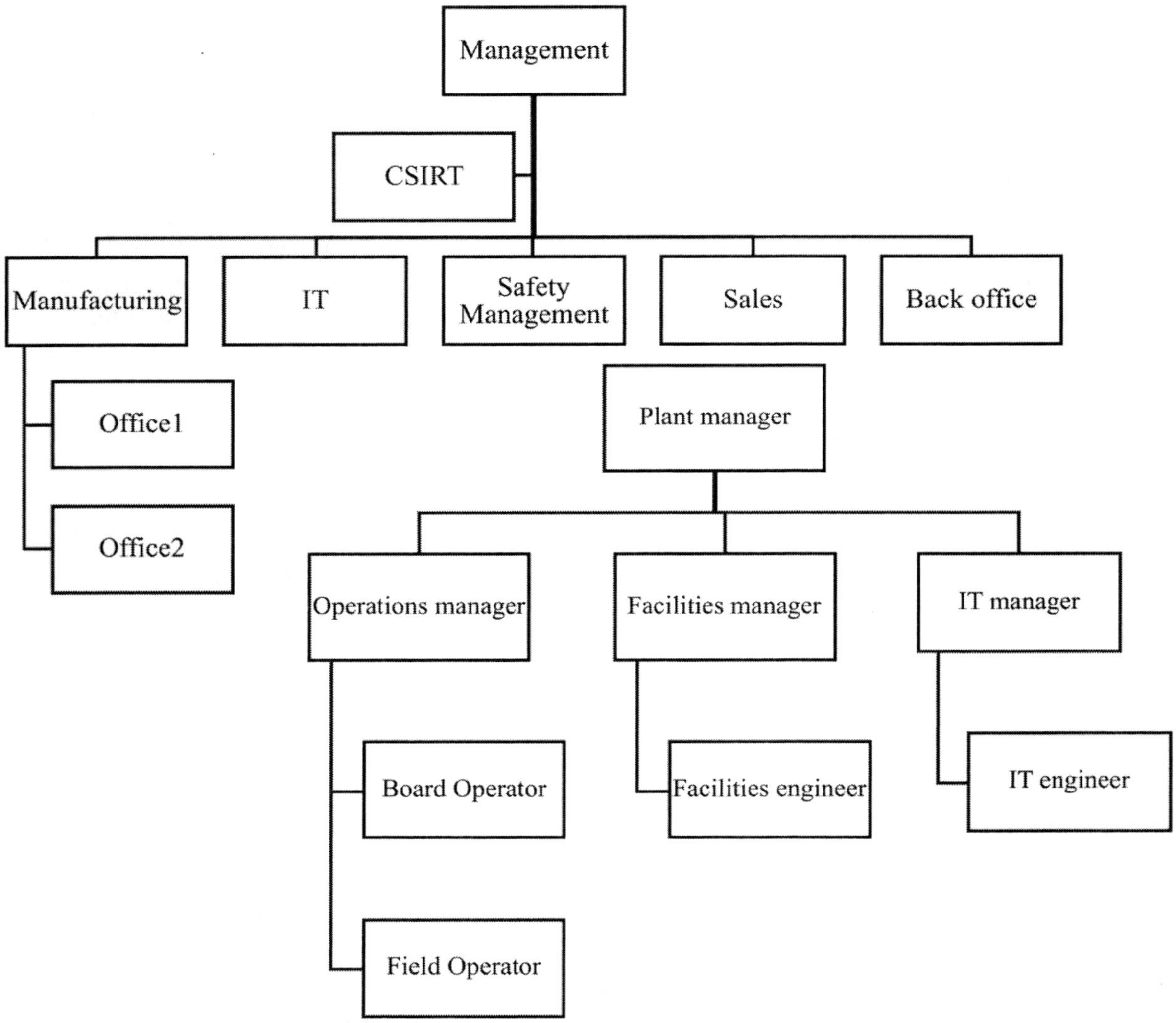

Figure 7. The organization chart of a virtual company

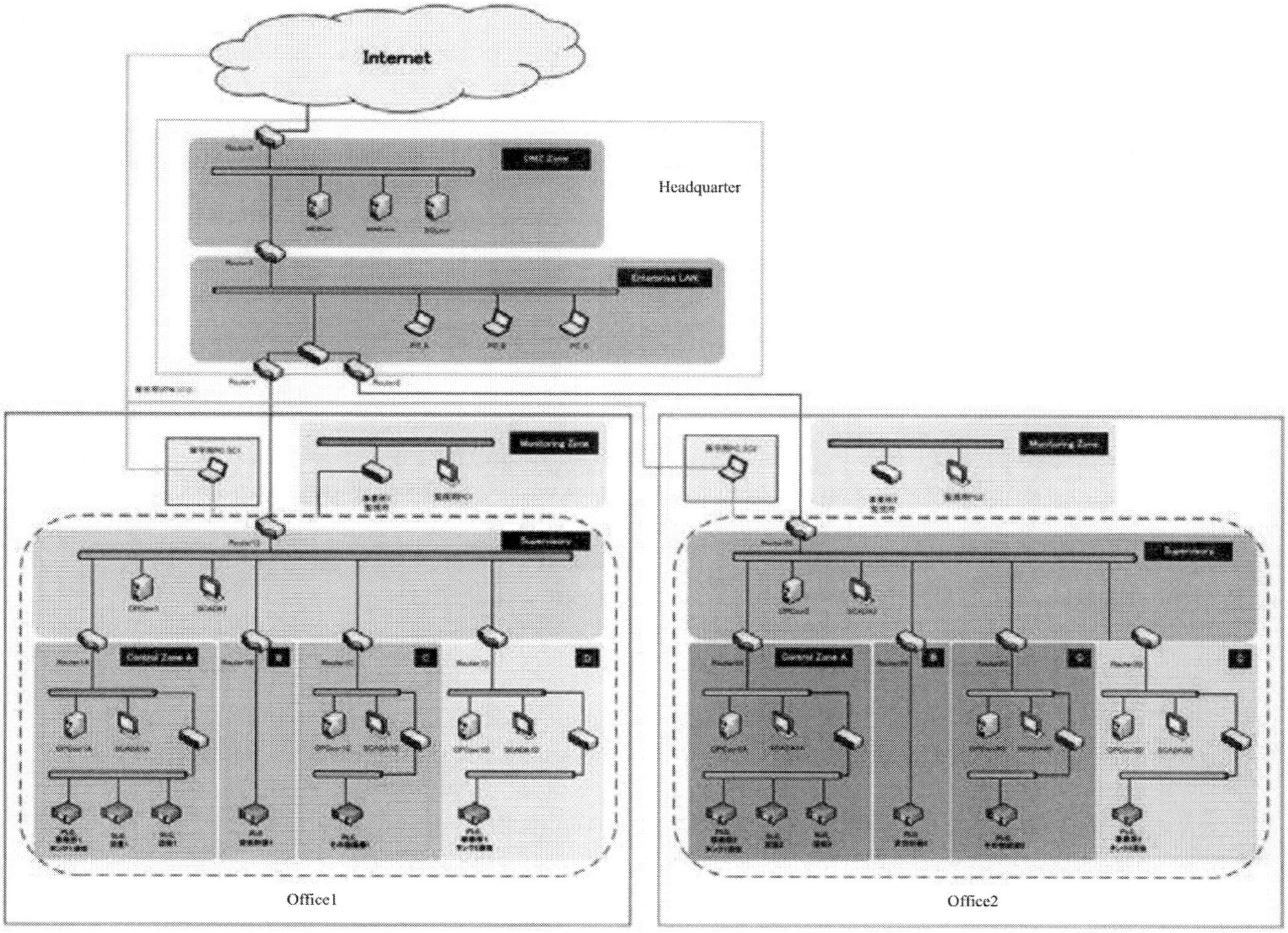

Figure 8. The network structure of the virtual company

Plant Operation Status Scenario:
A real-world plant consists of many devices, pipes, valves, and so on. If the plant is used as the subject of the exercise, the discussion may be complicated, and the purpose of the exercise may not be achieved. The plant used as the subject of the exercise should have a simple structure. By using a plant with a simple structure as the subject, it is possible to facilitate discussions focusing on the cyber incident that is the purpose of the exercise. In addition, it becomes easier to obtain meta-knowledge about cyber incidents, and it becomes easier to apply the knowledge gained through exercises to actual plants.

This scenario is based on our cybersecurity testbed. And we have introduced the control equipment used in the actual plant. This testbed aims to test and develop ICS security solutions. Based on the cyber-attack demonstration conducted as part of the research, the plant operation information caused by the cyber-attack is used for the exercise.

Cyber-attack Scenario:
The flow of cyber-attacks on plants is almost the same as cyber-attacks on IT systems.
To imagine how to respond to a cyber incident, it is important to deepen your cyber-attacks knowledge. There are many cyber-attacks methods, but attackers usually attack based on the cyber kill chain [8]. The cyber kill chain [9] classifies cyber-attacks into seven phases (Reconnaissance, Weaponization, Delivery, Exploitation, Installation, Command & Control, Actions on objectives). The idea is that appropriate measures can be taken in each phase, and cyber-attacks can be prevented if the attack does not proceed to the next phase.
This scenario is based on a cyber-attack demo conducted in our laboratory. It also includes the idea of a cyber kill chain. Throughout the exercises, participants can discuss the measures to protect the company by understanding cyber-attacks workflow and intention.

3.1 Card-Type Incident Response Exercise

The first exercise is Card-Type Incident Response Exercise. This exercise aims to gain an awareness of what a cyberattack on a plant is and what actions will be required.
This exercise uses a prepared action card to create a workflow for responding to cyber-attacks. Participants will achieve the purpose of the exercise by considering the response to cyber incidents while understanding the given action card's contents.

3.3.1 Exercise steps

These exercises consist of five steps: introduction, explanation of prerequisites, group work, presentations, and open discussions. Participants guess the company's situation from the given preconditions and create a workflow while thinking about the actions and information sharing necessary to make the plant safe and carry out business continuity.

Introduction:
The facilitator explains the purpose of the exercise and how to proceed. The general scenarios of the exercise are introduced to the participants. At this step, participants understand the importance of the exercise and, therefore, be motivated to apply themselves fully. By doing so, participants can increase the benefit of the exercise.

Description of Prerequisites:
Participants perform group work as a virtual company that owns a plant and provides products. The facilitator explains the preconditions about the virtual company at the beginning of the exercise. Prerequisites include the plant's status, the state of the IT network, and the company situation. Participants watch a video to understand the preconditions (In this phase, we use the videos we create to help participants understand the prerequisites.).

Participants envision the outcome based on the prerequisites.

Group Work:
The group task is to create a workflow by envisioning the outcome from the given information and considering the possible situation's response.
Originally, this exercise was conducted in group work in person. With COVID-19, we improved this exercise so that we could perform it using web conferencing.

In group work, the Excel file shown in Fig. 9 is used. Worksheets and action cards are pre-stored in this Excel file.

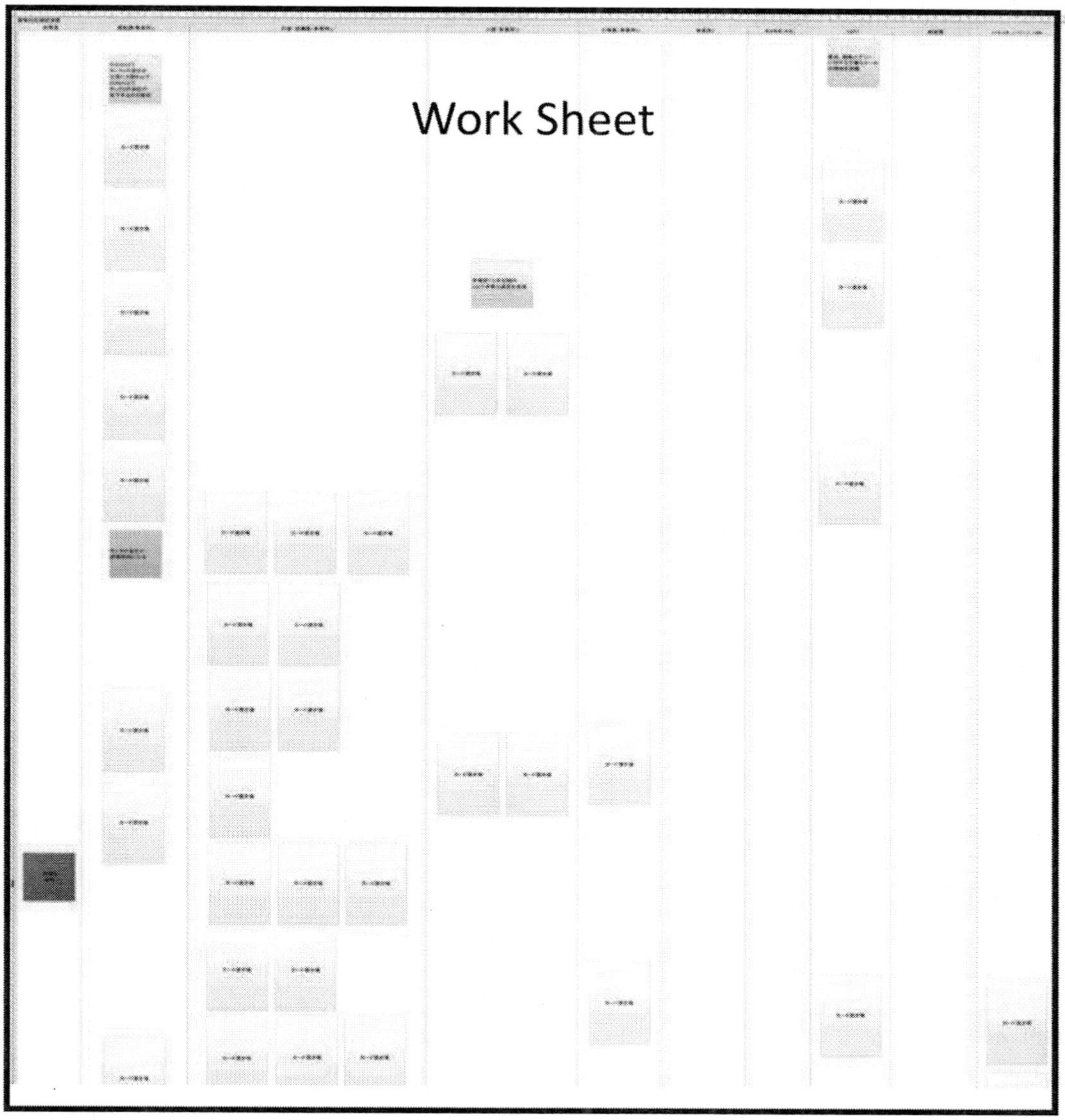

Figure 9. Group worksheet (Excel Screen)

The worksheet contains columns with the virtual company's departments, prerequisites, and action card storage roles. Action cards are classified into two types: normal cards and special cards (the upper part of Fig. 10 is the normal card and the lower part is the special card).

On-site Action — Action at headquarters — Plant shutdown/restart

Suspect a cyberattack — Information sharing

Figure 10. Type of action cards

The normal card is an incident response card, and there are a total of 38 cards of three types:

- "action at the site"
- "action at the head office"
- "shutdown/restart of the plant"

On the normal card, the result after taking action is written on the backside, and the participant thinks about the next action based on that information.

There are two types of special cards: "suspect/validation of cyber-attack" and "information sharing." The information-sharing card is used to share the information obtained by each department with other departments. In this exercise, participants discuss and strategize countermeasures from a bird's-eye-view, that is, without playing a specific role. However, in reality, information cannot be shared with others unless it is shared by the person who has the information. Participants use the information-sharing card to share the information held by each department. This makes it possible to express the flow of information sharing on the workflow. An important action in this exercise is to suspect/validate a cyberattack. No security response will be given unless the cause of the plant malfunction is suspected to be a cyber incident. There is information that can be a trigger to suspect that it is a cyber incident, and someone must judge that it is a cyber incident based on that information. By describing it on the workflow, participants identify the departments that will be the key persons in the cyber incident response. Participants will use these action cards to create a workflow, as shown in Figure 11.

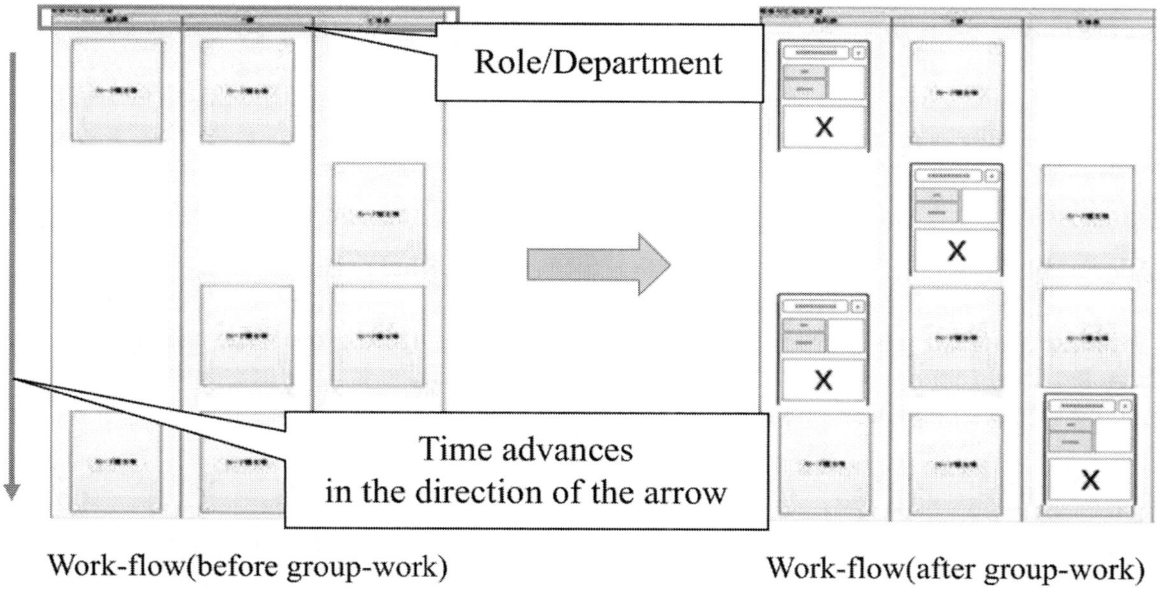

Figure 11. Workflow creation (image)

Presentation:
For groups with a lot of work experience in plants, create workflows that focus on plant safety. For groups with a lot of work experience in the IT department, create workflows that mainly support IT systems. Workflow, which is the product of group work, depends on group members' background and experience. By sharing the workflow and the workflow creation process, the exercise participants confirm the similarities and differences between the ideas of other groups and their own group's ideas. This step can enhance the effectiveness of the exercise.

When each group makes a presentation, ask them to focus on the following three points.

- How did you think about "who (which department), when, why" and placed the action card?

- What did you imagine when you created the workflow from the information given?

- How did you feel through the exercises what kind of human resources, information sharing, and organization-related systems should respond to cyber incidents?

Participants can understand the ideas of other groups and gain new knowledge by listening to other groups' presentations and asking and answering questions.

Open-Discussion:
After presenting the group's plans, the facilitator will show an example of the exerciser's workflow. This workflow presented is the optimal solution that the exercise developer thinks. The facilitator will give the same presentation as the participants in the exercise. After the

facilitator's presentation is over, the ideal inter-organizational collaboration for cyber incidents, information sharing, and necessary human resources will be discussed.

Through the exercises, exercise participants will experience the response to cyber incidents and understand the impact of cyber-attacks on ICS. And it is possible to gain insights on ideal organizational cooperation and necessary soft skills and hard skills so that the organization can respond quickly in the event of a cyber-attack.

3.4 ICS-BCP Creation Exercise

The second exercise is a BCP creation exercise. Participants create an incident response exercise based on a given scenario (such as a virtual business scenario). Participants will consider how to attack the plant from the attacker's point of view, consider actions to protect the company from cyber-attacks, and create an ICS-BCP. The purpose of the exercise is to understand the perspective of things needed to create an ICS-BCP.

3.4.1 Exercise steps

This exercise consists of five steps: scenario explanation, group work, demonstration, exercise review, and open discussion. Participants in the exercise identify the risks that the company faces from the given prerequisites. Next, consider an example of a cyber-attack that may cause that risk. Then, think about how to respond to the incident caused by the cyber-attack and create an ideal ICS-BCP.

Description of the Scenario:
Virtual enterprise scenarios, plant scenarios, and cyberattack scenarios are required to create incident response exercises. In this exercise, we want participants to discuss how to create an ICS-BCP rather than how to create a scenario. Therefore, the scenario is provided by the facilitator in advance. In addition, since there id an infinite number of attack scenarios, some scenarios are presented, and participants select from those scenarios. In this step, those scenarios will be explained.

Group-Work:
This exercise also creates deliverables through group work.
The deliverables to be created are the worksheets and action cards required for the card-type incident response exercise.

- First, participants analyze the risks of a virtual company from a given virtual company and plant scenario.
- Next, participants create a cyber-attack scenario that causes the analyzed risk. Participants create a scenario of the impact on the company and the plant's operating status due to the created cyber-attack scenario.
- Finally, participants create an ideal workflow for the company to perform incident response exercises and create actions that use that workflow as action cards.

Figure 12. An example of Team Discussion in Group Work

Demonstration:
In this step, participants perform their own exercises on other groups. Exercises will be conducted for different groups using workflows and action cards created in group work. Through this step, participants confirmed that they could convey the message through the exercise (ideal response / inter-organizational collaboration) to other groups. Participants can also gain more cyber-incident experience by taking exercises in various scenarios.

Group-work (Review of the exercise):
There are many things that can be gained by doing exercises.
Participants review the exercises they have created based on the results of the exercises conducted by other groups. The group once again discusses whether the ideal workflow that they have considered is the optimal solution, what factors hinder the ideal workflow that they have created, and so on.

Open-Discussion:
Each group will share what they have reviewed their exercises and what they have learned through the exercises in this step. By listening to various people's opinions through this step, it is possible to discuss how to view the things necessary for creating an ICS-BCP.

3.5 IMANE (Simulation exercise)

This exercise aims to confirm the effectiveness of the created ICS-BCP by simulating the incident response. This exercise is a simulation exercise that uses a personal computer to create a game based on a scenario, input chat, and respond to an incident by selecting a response.

3.5.1 Exercise steps

This exercise consists of five steps: identifying stress factors such as no one to deal with or no tools to deal with, adding storylines, simulating incident response, reviewing the exercise, and open discussion. Participants will play the role given and aim for business continuity while responding to incidents as virtual company members.

Identifying Stress Factors:
The ICS-BCP created by ICS-BCP creation exercise is an ideal workflow. In other words, the plan is based on the premise that all the assumed actions can be performed. However, in reality, it may not be possible to perform actions that should be possible due to various factors, such as lack of resources such as people and tools required to perform actions. Participants will identify stress factors that can hinder their actions in operating the ICS-BCP they have created. Stress factors are the risks involved in conducting ICS-BCP, and it is important to have advanced preparations and alternatives to eliminate the stress factors.

Adding storylines:
The created ICS-BCP only identifies the actions and the shared information, instruction contents, and information on the operation transition of the plant due to the actions are hidden between the actions. Participants identify the hidden information and necessary information communication paths. Then, create a game scenario by incorporating the stress factors created in the previous step.

Simulating Incident Response:
This step simulates an incident response. Figure 13 shows the structure of this exercise. In previous exercises, exercise participants were required to create a workflow in groups without being assigned to a specific role. In this exercise, roles are assigned to each exercise participant, and incident response is performed by taking actions for each role. It is an exercise that is closer to the actual incident response.

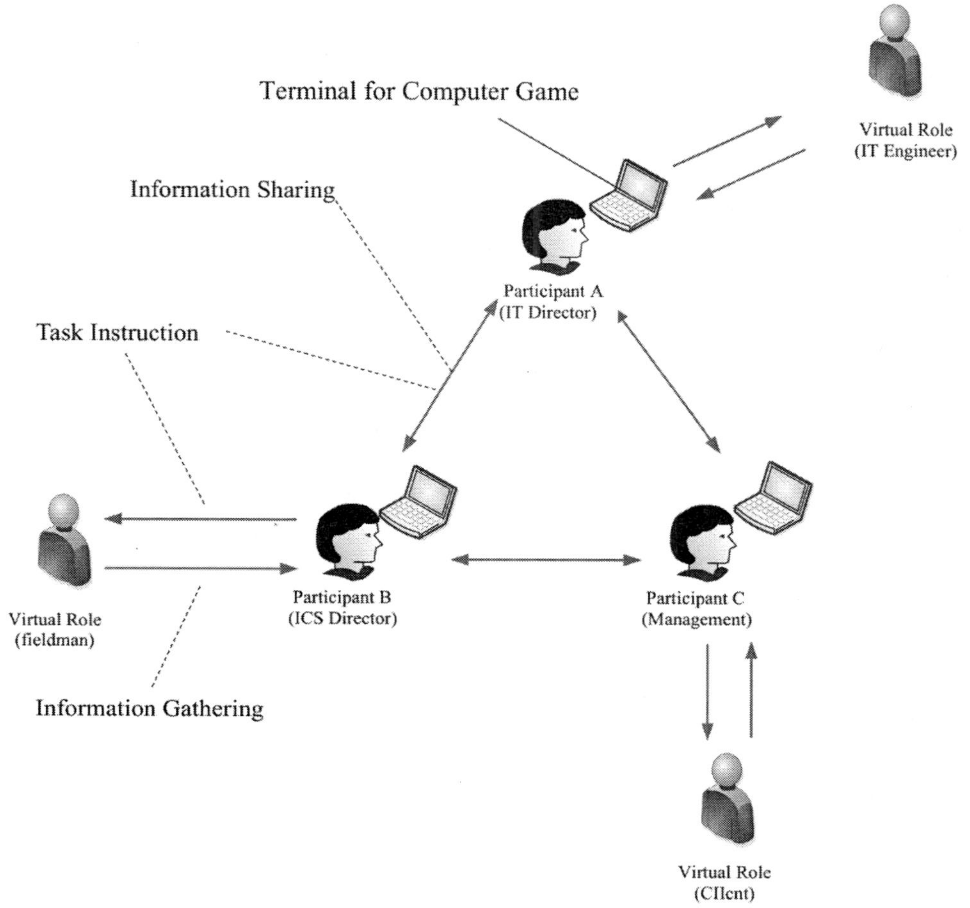

Figure 13. IMANE structure

Figure14 is a console screen operated by the participants. Participants share information with other players and take actions based on the information flowing in the tray. Each participant is required to understand the situation and take action based on the information displayed on their console screen. In addition, since the actions that can be performed for each role and the information is given are different, it is also required to share the information necessary for taking action to converge the situation with other players. Participants use the interface in Figure 14 to share information and select actions. The information displayed in the tray in Figure 15 is linked to the story created in the previous step, and the information displayed differs depending on the action performed so that the incident response can be made more realistic.

Figure 14. The user interface of IMANE exercise

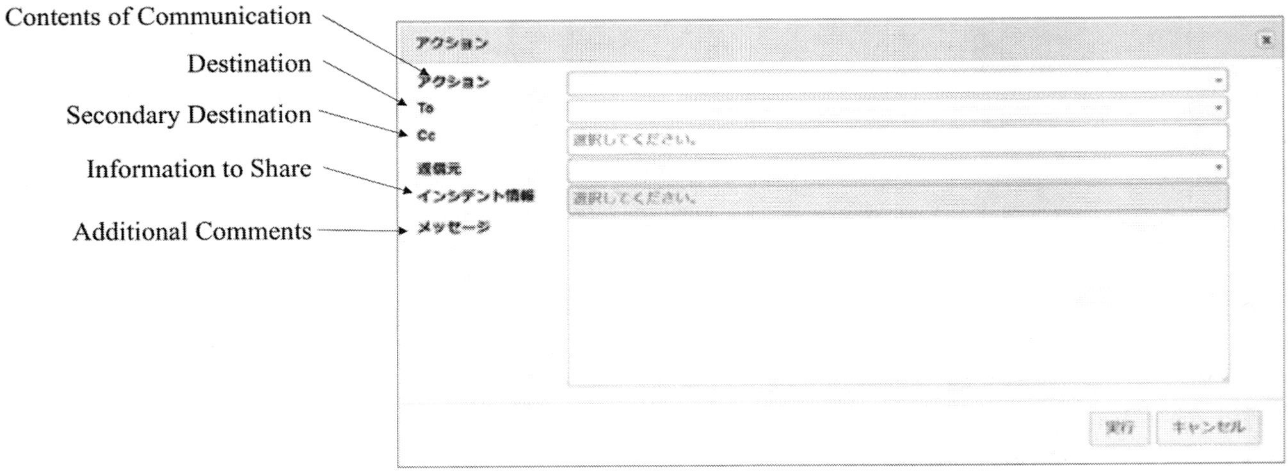

Figure 15. The action user interface of IMANE Exercise

Review of the Exercise:
After the end of this exercise, it can be output as a product of the exercise, as shown in Figure 16. This workflow includes the information given to each role, the actions taken, and the flow of information-sharing and participants can use this workflow to review the incident response after the exercise.

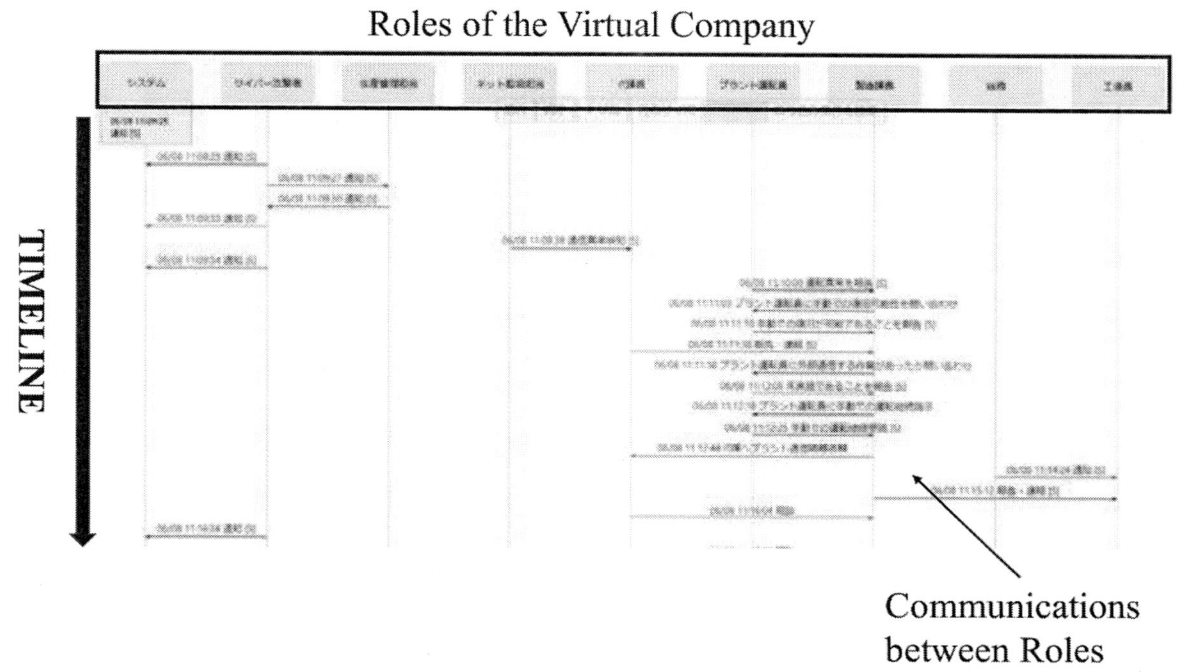

Figure 16. Example deliverable of the exercise

Using this deliverable, exercise participants can objectively review ICS-BCP by looking back on their response. It is possible to improve the accuracy of ICS-BCP by checking for missing or defective ICS-BCP.

Open-Discussion:
In this step, each participant shares the impressions gained through the exercise. We discuss the ideas needed to create an ICS-BCP and human resource development using this exercise.

By performing these exercises in various scenarios, the exercise participants can increase their imagination and resilience to cyber incidents.

4 Pilot Exercise

While improving the developed exercises step by step, many security and operation experts participated in the Nagoya Institute of Technology security workshops from 2018. To get an opinion to improve the exercises, we asked the participants to complete the questionnaire. The questionnaire was answered with a maximum of 5 points. In the exercise conducted on December 11, 2020, 28 people participated, and the following questionnaire results were obtained (Max point is 140). From the results of this questionnaire, it can be seen that the people who participated in the exercise understood the purpose of the exercise that we set and received high praise for the exercise.

Questionnaire content for Card-Type Incident Response Exercise and ICS-BCP Creation Exercise:

1. Was the goal set for the exercise, and was it possible to attain it?
2. Do you think you learned about organizational collaboration in the event of a cyber incident?
3. Do you think you learned about the organizations needed to respond to cyber incidents?
4. Was the information necessary to proceed with the exercise given?

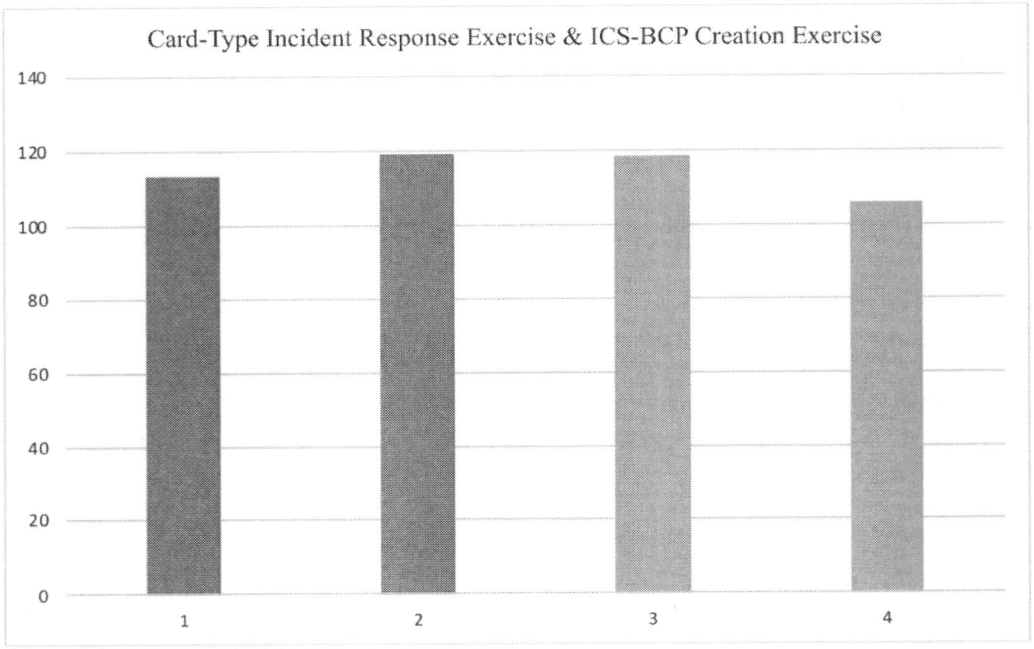

Figure 17. Questionnaire results of Card-Type Incident Response Exercise & ICS-BCP Creation Exercise

Questionnaire content for IMANE:

1. Was the goal set for the exercise, and was it possible to get it?
2. Do you think you learned about organizational collaboration in the event of a cyber incident?
3. Do you think you learned about the organizations needed to respond to cyber incidents?
4. Is it designed so that necessary information is communicated and given to each role?
5. Do you feel that you were able to experience information gathering and decision-making in an emergency?

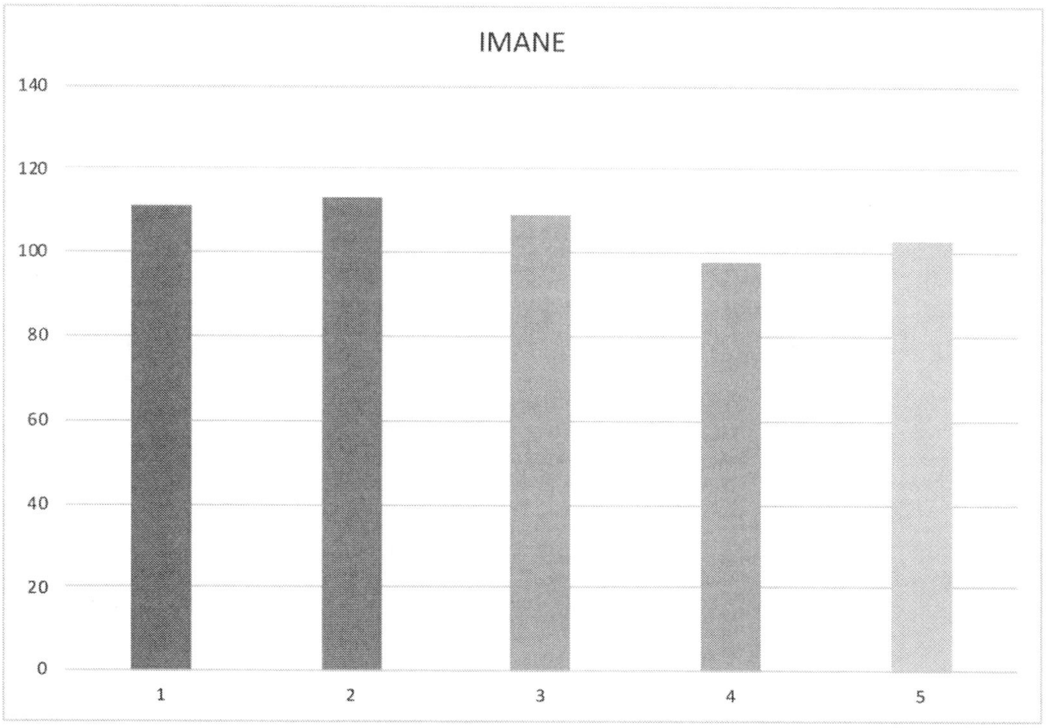

Figure 18. Questionnaire results of IMANE

In addition, the following was found by analyzing the deliverables of the exercise.

- Plant safety first

Many workflows can keep the plant safe, even if the backgrounds of the participants in the exercises vary. Even if the cause of the plant abnormality is a cyber-attack, it is easy to imagine the visible situation, and it can be seen that appropriate measures can be taken.

- No action has been taken to identify a cyber-attack

Analyzing the workflow, it is possible to select situation sharing and actions for performing Safety Response. However, information sharing for performing Security Response, such as identifying the range of influence of cyber-attacks is often not successful. This is thought to be due to the fact that they have never experienced cyber-attacks and lack the knowledge to consider actions. In addition, Safety Response is easy to imagine because it corresponds to the visible physical impact on the plant, but Security Response is difficult to imagine because it corresponds to the invisible network.

- Organization that is the center of the response

To respond to cyber incidents to ICS, it is necessary to fluidly change the organization that is the core of the response.
In the initial response, it is necessary to respond to plant operation abnormalities caused by cyber incidents. The person working at the plant takes action. After ensuring the plant's safety, it is necessary to eliminate the effects of cyber incidents. IT department takes this action. Finally, it is necessary to continue the business considering the impact of the cyber incident on the business.

Management considers actions for the business gains.
In this way, the work to be done changes over time, and it is necessary to respond while changing the organizational response that is the core of the response accordingly.

5 Concluding Remarks

In this paper, the authors presented the factors required of a company against the threat of cyber-attacks targeting ICS and explained the exercises developed to increase the company's resilience.

Through the workshop, we received positive opinions from the participants that they could gain various awareness of cyber-attacks. However, the effect of the exercise could not be confirmed quantitatively, and it could not be confirmed that the response-ability was improved.

In the future, we will conduct exercises for more people and obtain deliverables. Then, by analyzing the deliverables, we will consider the method of planning ICS-BCP considering cyber-attacks and the conditions necessary to confirm the validity of the planned ICS-BCP.

6 References

[1] Institute for Science and International Security, "Did Stuxnet Take Out 1,000 Centrifuges at the Natanz Enrichment Plant?", https://isis-online.org/isis-reports/detail/did-stuxnet-take-out-1000-centrifuges-at-the-Natanz-enrichment-plant/8 (accessed Jan. 25, 2021)

[2] ICS-CERT,
ICS Alert(IR-ALERT-H-16-056-01), "Cyber-Attack Against Ukrainian Critical Infrastructure", https://us-cert.cisa.gov/ics/alerts/IR-ALERT-H-16-056-01 (accessed Jan. 25, 2021)

[3] SANS Institute, ICS Defense Use Case No. 6, "Modular ICS Malware", https://ics.sans.org/media/E-ISAC_SANS_Ukraine_DUC_6.pdf (accessed Jan. 25, 2021)

[4] Norsk Hydro, "Cyber-attack on Hydro", https://www.hydro.com/en/media/on-the-agenda/cyber-attack/ (accessed Jan. 25, 2021)

[5] BSI, "Industrial Control System Security Top 10 Threats and Countermeasures2019", https://www.allianz-fuer-cybersicherheit.de/ACS/DE/_/downloads/BSI-CS/BSI-CS_005E.html#download=1 (accessed Jan.25, 2021)

[6] T. Aoyama, M. Koike, I. Koshijima, Y. Hashimoto, "A unified framework for safety and security assessment in critical infrastructures". In: Safety and Security Engineering V. WITPRESS LTD. (Sep 2013)

[7] Y. Ota, T. Aoyama, D. Nyambayar, I. Koshijima, "Cyber Incident Exercise for Safety Protection in Critical Infrastructure," Int J. Saf. Secur. Eng. 8,246-257 (2018)

[8] J. Sakhnini, H. Karimipour, A. Dehghantanha, R.M. Parizi, G. Srivastava, Security aspects of internet of things aided smart grids: a bibliometric survey. Internet Things (2019). https://doi.org/10.1016/j.iot.2019.100111

[9] A. Shostack, *Threat Modeling: Designing for Security* (Wiley, Indianapolis, In, 2014)

Hazard Identification during Process Related

Maintenance and Operations Tasks

T Michael O'Connor
Mary Kay O'Connor Process Safety Center
Texas A&M University, College Station, TX
mikeocon@gmail.com

Dr Zohra Salim
Mary Kay O'Connor Process Safety Center
Texas A&M University, College Station, TX
zohrahalim@tamu.edu

Prepared for Presentation at
American Institute of Chemical Engineers
2021 Spring Meeting and 17th Global Congress on Process Safety
April 18 - 23, 2021

Hazard Identification during Process Related Maintenance and Operations Tasks

T Michael O'Connor
Mary Kay O'Connor Process Safety Center
Texas A&M University
College Station, TX
mikeocon@gmail.com

Dr Zohra Halim
Mary Kay O'Connor Process Safety Center

Keywords: serious injury, fatality, chemical manufacturing, petroleum refining

Abstract

Trends in process related fatalities are presented for chemical manufacturing and petroleum refining. The trend analysis revealed the more than half of fatalities occur in single fatality incidents. This observation led to study of the nature of these low fatality incidents. Three common tasks were identified that are associated with a large portion of process related fatalities and severe injuries. These 3 tasks are clearing, cleaning and purging lines and vessels; opening lines and vessels; and transferring fluids. The specific hazards encountered in incidents associated with these tasks are identified. The goal is to provide the basis to improve system design, training, procedures, job hazard analysis, PPE requirements and other factors based on hazards actually associated with major harm to individuals performing these tasks.

Introduction

There are 3 primary components of this study:

- Trends in fatalities from 1984-2019 in chemical manufacturing (CM) and petroleum refining (PR)
- Tasks associated with severe injuries and fatalities in CM and PR
- Nature of serious injury incidents in CM only

Data is extracted from two US OSHA databases:

The first is the Fatality and Catastrophe Inspection Search (FCIS) database which includes all incidents with a fatality or 3 or more injuries. Since incidents without a fatality and less than 3 injuries are not included in the data injuries from this database are not included in the study. This database covers the years 1984 to the present whichallows long-term trending including the implementation of OSHA PSM in 1992. This analysis is limited to the states in OSHA Region 6. These states are TX, LA, AR, OK and NM. The data was accessed through OSHA's online query system.

The second database is the Severe Injury Reports (SIR) database. This database is limited to the 28 States with federal, not state, OSHA programs. This database includes all the Region 6 states. It covers the years 2015 to July 2020 and includes cases with hospitalization or amputation. This entire dataset can be downloaded as a spreadsheet.

For both databases the authors have interpreted the accident description to determine whether it is a process related incident and the task being undertaken. The serious injury database contains the hazardous substance, nature of injury, affected body part and event type. In some cases, the description is not sufficient to extract the needed data and these items are classified as unknown. Some incidents have been investigated by the Chemical Safety and Hazard Investigation Board (CSB) or other agencies. In those cases, information may have been extracted from those sources.

This paper focuses on both the Chemical Manufacturing (CM) and Petroleum Refining (PR) industry as determined from their NAICS and SIC codes. Both industries were analyzed for fatality trends from 1984-2020. The tasks associated with both severe injuries and fatalities were determined for both CM and PR. A more detailed analysis of selected tasks was performed only for CM. Detailed analysis of PR incidents are not yet completed.

Scope of Incidents in Study

The following definitions were utilized to determine if an incident was process related and was to be included in the study:

Hazardous Material – Any substance the is toxic, reactive, flammable, asphyxiating or that presents a hazard to people, the environment, or property due to pressure or temperature.

Process Safety Incident – The sudden, unintended release of, or exposure to, a hazardous material, associated with an industrial process, that results in or might reasonably have resulted in death, injury, significant environmental or property damage, evacuation or sheltering in place.

These are generic definitions used by the MKOPSC. In this paper all of the incidents involve a fatality or injury since they are from the OSHA databases.

Trends in Fatalities in Chemical Manufacturing and Petroleum Refining

Long term trends with time are available from the Inspection Database for fatalities. Serious injuries are included in the database if there are 3 or more injuries or they occur with a fatality. However, incidents with one or two injuries are not included making any trending information questionable. The Serious Injury database only covers the years 2015 through 2020, making any trends of doubtful significance, so only fatalities were used for trending.

Plots are shown for CM and PR. Fatalities and fatal incidents over the years 1984 to 2020 are illustrated. There are a number of high fatality incidents in the latter 1980s. This is followed by a dramatic decrease in fatalities around the time of the implementation of OSHA PSM in 1992. It seems likely this decrease was due to both the increased focus on process safety due to the new regulation as well as a direct response to these high fatality incidents. The trends are somewhat similar for CM and PR with the exception of the BP Texas City explosion in 2005.

Figures 1 through 4 show the large number of incidents with a low number of fatalities. After 1992, for CM and PR, greater than 80% of fatalities occurred in single fatality incidents. This statistic in particular motivated a more detailed examination of these numerous incidents with mostly low numbers of fatalities as well as incidents with serious injuries.

In this plot (see Figure 1), the total height of the bars represents the total fatalities in a year. The colored bands making up the bars show individuals incidents and the number of fatalities in that event. The red arrow denotes the implementation of OSHA PSM in 1992. Clearly there is a sharp decrease in fatalities after that time. In addition, there are no incidents with more than 4 fatalities after PSM began.

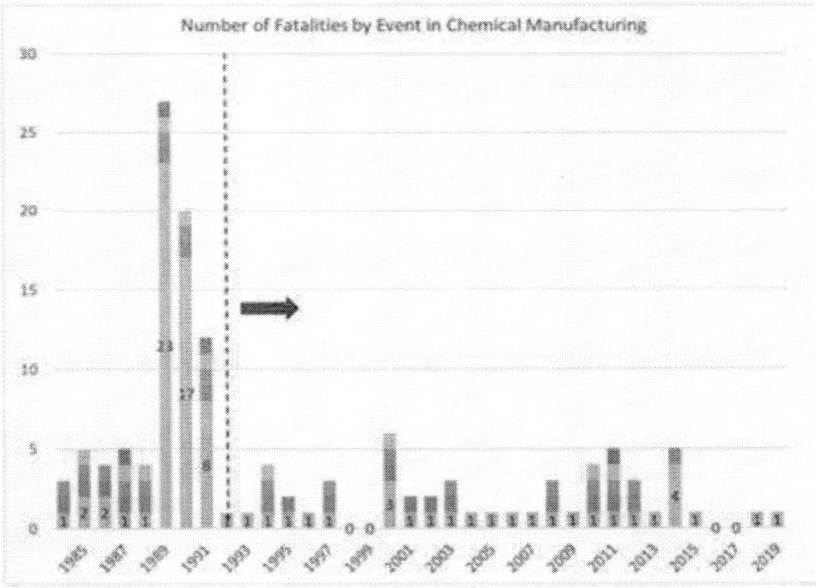

Figure 1 - Fatalities by Event in Chemical Manufacturing Region 6

This plot (see Figure 2) shows that not only the annual number of fatalities dropped in 1992 but also the number of fatal events.

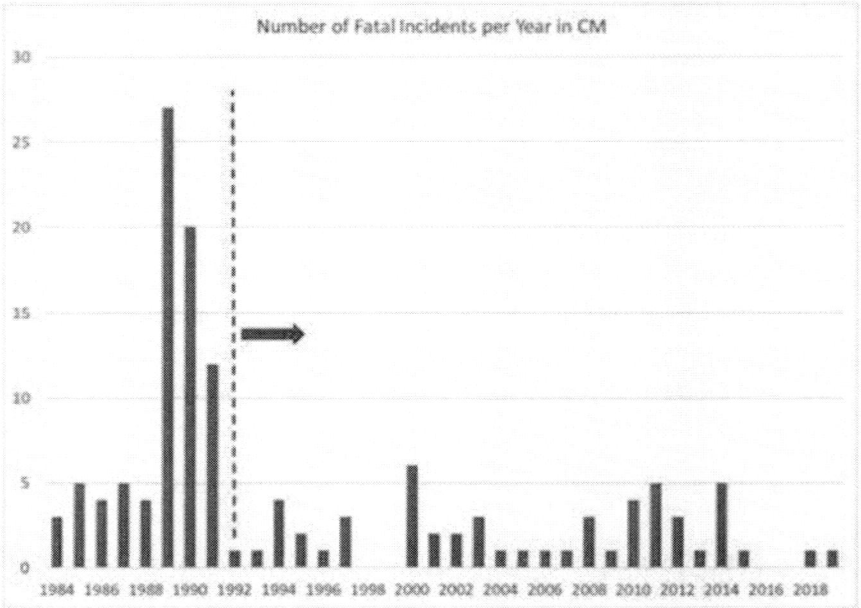

Figure 2 - Fatal Events in Chemical Manufacturing

Like CM there is a distinct reduction in fatalities in PR about the time of PSM implementation. This reduction appears to have occurred in 1994, a few years after PSM. There is another reduction about the time of the event at BP Texas City (see Figure 3). We theorize that this high-profile event may have re-focused the PR industry on process safety. This reduction has persisted for 14 years.

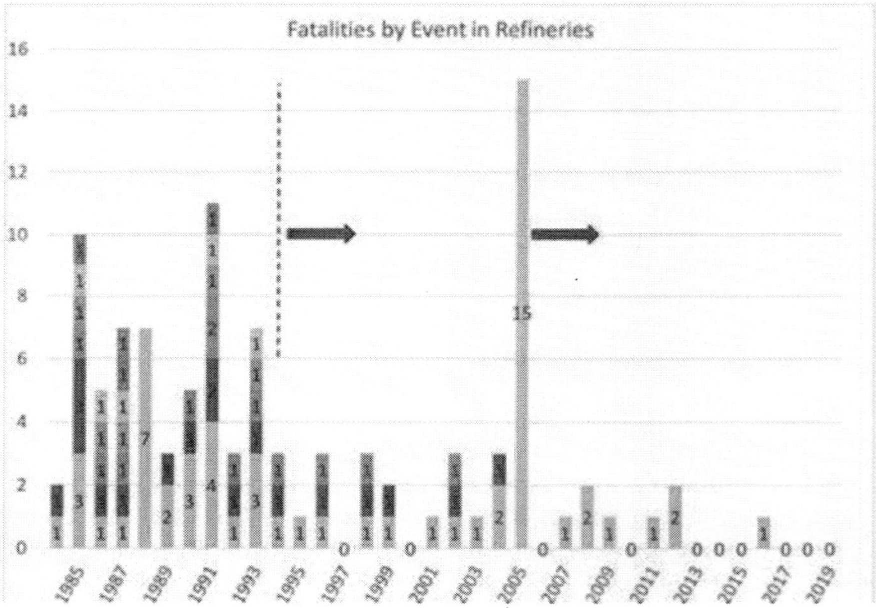

Figure 3 - Fatalities by Event in Petroleum Refining

It is clear that not only the number of fatalities was reduced around 1994 but also the number of events. Since 2002 there has been either zero or one fatal event per year (see Figure 4).

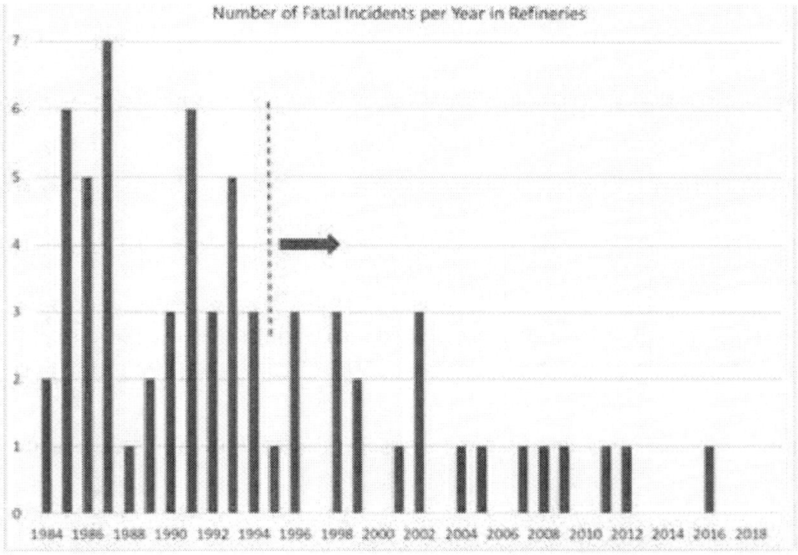

Figure 4 - Fatal Incidents in Petroleum Refining

Table 1

The fatality and fatal event annual rates are shown for CM and PR in Table 1. For CM there is a 50% reduction in fatal events per year after 1992. The fatalities per year dropped dramatically by a factor of almost 5. For PR there is a 60% reduction in fatal events in 1994 and an overall reduction of 75% after 2005. Fatalities in PR were reduced by nearly 50% in 1994 and a remarkable 10 to 1 after 2005. The relative magnitudes of the PR and CM rates are not comparable since they are for different numbers of facilities and employees.

Event and Fatality Rates for CM and PR - Region 6			
Chemical Manufacturing			
	1984-1992	**1993-2020**	
Fatal Events/yr	**4.1**	**2.0**	
Fatalities/yr	**10.0**	**2.2**	
Petroleum Refining			
	1984-1994	**1995-2005**	**2006-2020**
Fatal Events/yr	**4.0**	**1.6**	**1.0**
Fatalities/yr	**6.0**	**3.2**	**0.6**

Tasks Associated with Serious Injuries and Fatalities

As the descriptions of each incident were analyzed it was soon apparent that most of the incidents involved a task being performed by a small number of people and the consequences typically suffered by those workers. Both maintenance and operational activities are involved.

Table 2 lists the tasks for serious injury and fatal incidents, from both databases, in both CM and PR. The term "unrelated" is used to denote incidents that are not related to the activity of the victim. Most of these "unrelated" cases are mechanical failures, leaks or explosions and the victim is by chance in the vicinity.

Table 2

Severe Injury and Fatal Incidents by Task									
	Chemical Manufacturing				Petroleum Manufacturing				
	Severe Injury Incidents	%	Fatal Incidents	%	Severe Injury Incidents	%	Fatal Incidents	%	% Averaged for Sort
Cleaning, clearing, purging lines and vessels	35	16	16	20	10	29	11	11	19
Opening lines and vessels	47	21	8	10	3	9	15	15	14
Transferring fluids	43	20	8	10	3	9	7	7	11
TOTAL % for 3 Tasks		57		39		47		32	
Unrelated*	10	5	11	13			16	16	8
Unknown	20	9	3	4	2	6	8	8	7
Maintenance	18	8	3	4	5	15	4	4	8
Operating	6	3	6	7	3	9	4	4	6
Starting equipment	6	3	5	6	2	6	5	5	5
Confined Space Entry			7	9			8	8	4
Mixing chemicals	8	4	3	4			7	7	4
Troubleshooting	7	3			2	6	4	4	3
Fall or step into hazard	3	1			2	6	1	1	2
Other	5	2	4	5			1	1	2
Sampling	3	1	3	4	1	3			2
Handling explosives			3	4					1
Lighting heater							5	5	1
Transferring solid	7	3							1
Reactor startup			2	2					1
Starting process					1	3	2	2	1
Hot Work							3	3	1
Testing	2	1					1	1	0
Inspecting							1	1	0
Total Incidents	220	100	82	100	34	100	103	100	100

*Unrelated indicates the employee was not involved in a task telated to the event.

Tasks Associated with Serious Injury and Fatal Incidents in Chemical Manufacturing.

In both CM and PR, there are 3 activities that account for a large fraction of the fatalities. These tasks are:

- Cleaning, clearing, or purging lines and vessels
- Opening lines and vessels
- Transferring fluids and solids

These 3 types of tasks usually involve the intentional interruption of a pressure containing equipment component. That is of course the case with opening lines and vessels. For cleaning, clearing, or purging, it is necessary to open the pressure containment before proceeding. Transferring fluids often involves making or breaking connections as well as venting and draining lines and vessels

Cleaning, clearing, and purging by their nature involve removal of a substance, usually from the interior of the containment. Difficulties are encountered in numerous ways. These include:

- Substance is not what was expected
- Quantity of substance is greater than expected
- Unexpected pressure behind a blockage
- Blockages are propelled out of the containment by pressure
- Tools, covers and hatches are propelled by unexpected pressure
- Containment is over-pressurized when clearing an obstruction

Opening lines and vessels often involve problems similar to those above. These include:

- Unexpected pressure
- Unexpected substance
  Unexpected quantity

Transferring fluids involve many different situations based on the source of the fluid, receiving container, and type of transfer line or hose.

- Hoses often disconnect unexpectedly
- Hoses are subject to failure
- Disconnected hoses may contain an unexpected substance
- Hard pipes often contain vents and drain lines that are open
- Transfers may result in mixing incompatible substances
- Rail cars and tank trucks may have complex sets of valves for venting, draining, multiple compartments and other confusing situations
- Totes, drums, buckets, and other containers can be dropped or damaged, sometimes by forklifts
- Static electricity may be an issue in many of these operations

Nature of the Serious Injury Incidents in CM

The OSHA database, SIR, contains four factors that describe the incident and the resulting injuries. The four factors are the following:

- Type of Event
- Hazardous Substance
- Nature of Injury
- Affected Body Part

The results for all tasks in CM are shown in four tables. These four factors are the results of the incident and infer a limited amount about the causes. Hopefully information about the most common hazards will assist in focusing efforts for improved safety. Data for all incidents are included, not just those for the 3 most common tasks.

Table 3

Chemical Manufacturing		
Serious Injury - All Tasks		
Event Type	#	%
Exposure through intact skin, eyes, or other exposed tissue	82	37
Contact with hot objects or substances	74	33
Inhalation of harmful substance-single episode	21	9
Ignition of vapors, gases, or liquids	14	6
Struck by dislodged flying object, particle	9	4
Explosion of pressure vessel, piping, or tire	8	4
Explosion, n.e.c.	6	3
Other	9	4
TOTAL	223	100

Event Type

Two types of events account for 70% of all events in roughly equal frequency. These two are exposure of external tissue to a hazardous substance and contact with hot objects and surfaces. These are such broad categories that it does little to identify a particular type of event which deserves more attention than others. Two other event types also occur in a significant portion of incidents. These are inhalation of a substance and ignition of a substance. In planning for and executing a task, it is important to consider if any of these events could occur. This can lead to identifying procedures to avoid the event and to utilize proper PPE and emergency procedures.

Table 4

Hazardous Substance

Perhaps surprisingly, water is the most frequent hazardous substance involved in serious injuries. Of course, this is due in part to its widespread presence in many processes. Often the water is condensate and results in burns. Nearly as common are acids followed by resins. Resins are often associated with processes such as extrusion were contact burns are common. While the table shows a number of common substances there are an additional 40 substances that were only involved in one incident. In addition, there are 30 objects. In OSHA's classification system, objects, rather than a hazardous substance, are identified as the source of injury. These are usually cases where an object propelled by pressure or an explosion strikes the victim. These objects are typically shrapnel, pipe plugs, covers and hatches or hoses.

Chemical Manufacturing		
Serious Injuries - All tasks		
Substances	**#**	**%**
Water	**27**	**12**
Unknown, not classified	27	12
Acids	**25**	**11**
Resins	**17**	**7**
Multiple	12	5
Chlorine, chlorine bleach, compounds	**9**	**4**
Steam and vapors	**9**	**4**
Sodium and potassium hydroxide, potassium carbonate	**8**	**4**
Ammonia, anhydrous ammonia	**6**	**3**
Chemicals, n.e.c.	6	3
Phenol	**4**	**2**
Liquids, unspecified	4	2
Aromatics and hydrocarbon derivatives, except halogenated, n.e.c.	**3**	**1**
Other Substances	40	18
Physical Objects*	30	13
TOTAL	**227**	**100**
* OSHA lists objects such as those propelled by pressure		

Table 5

Nature of Injury

The nature of injury generally reflects the type of event shown in the previous table. Thermal burns are naturally associated with contact with hot objects and substances and chemical burns with exposure of tissue. Traumatic injuries result from being struck by object and explosions. Poisoning, toxic, noxious, or allergenic reactions are typically due to inhalation. While there is little information in these databases about PPE it appears that it is often inadequate for the actual circumstances.

Chemical Maufacturing		
Serious Injuries - All Tasks		
Nature of Injury	**#**	**%**
Thermal Burns	**93**	**40**
Chemical Burns	**79**	**34**
Traumatic	**26**	**11**
Poisoning, toxic, noxious, or allergenic effect	**26**	**11**
Other	**9**	**4**
TOTAL	**233**	**100**

Table 6

Affected Body Parts

One third of the incidents involve multiple body parts. This may reflect the serious nature of the studied incidents. Severe injuries may be the result of releases of larger quantities of hazardous substance, higher pressures and other factors that lead to more widespread injuries. This contrasts with less serious injuries associated with more common cases involving days away from work. In a study of these injuries (Mannan), there were typically specific body parts injured. The most often injured body part was the face. The second most affected body part for severe injuries is body systems. This means internal systems implying inhalation or ingestion of a substance. The remaining body parts are more typical parts of face, extremities, trunk, and head. The adequacy of PPE may be a factor in the severity of these injuries.

Chemical Manufacturing		
Serious Injuries - All Tasks		
Body Parts	#	%
Multiple body parts, n.e.c.	77	33
Body Systems	28	12
Nonclassifiable	28	12
Face	25	11
Lower Extremities	25	11
Upper Extremities	24	10
Trunk	15	7
Trunk & Head	8	3
TOTAL	230	100

Incidents with Detailed Investigations

The CSB and the US Department of Labor investigated several of the CM incidents in this study. These incidents were Phillips 66 1989, Honeywell 2003 (3 incidents) and DuPont 2014. These incidents generally involve one of the 3 common tasks. Phillips 66 involved the removal of a plug in the settling leg of a reactor. This led to a release of isobutane resulting in 23 fatalities and more than 100 injuries. Honeywell had 3 incidents involving 3 different substances. The first was a chlorine release due to a mechanical failure. The second was an antimony pentachloride release which resulted in a fatality. The release was the result of removing a plug from a cylinder. The third release was hydrogen fluoride during the transfer of fluid using a venturi stick, resulting in hospitalization of one worker. The DuPont incident resulted in 4 fatalities during the removal of a hydrate plug, releasing mercaptan. These three investigation reports show that larger scale incidents may be the result of the 3 common tasks.

Observations

The trends in fatalities before and after the implementation of PSM in 1992 shows a remarkable improvement in fatalities and fatal events. Following the reduction in 1992, most fatalities occur in single fatality incidents. These trends are counter to what one might conclude by observing high profile incidents in the media. This is true in part because of incidents in storage, transportation and at user facilities which are not included in this study.

The nature of these incidents seems to straddle the line between process safety and occupational safety. It may be that there is a need to better coordinate the safety efforts of what are in many companies two distinct organizations. The same may be true of the educational system. For example at Texas A&M University, occupational safety is taught in the School of Public Health while process safety is addressed in the College of Engineering.

Based on the incident description, the vast majority of fatalities appear to occur with relatively small releases affecting workers in close proximity to the release. At least 80% of the events are associated with what seem to be routine maintenance and operations tasks. They do not appear to be the result of complex technical failures such as impaired control systems.

There has been great success in limiting fatalities since the introduction of OSHA PSM. In order to further reduce fatalities, it appears that additional focus on less complex small scale releases is warranted.

The nature of these incidents suggest that improvements may be needed in training for hazard recognition, procedures, and job safety analysis.

The incidents are often caused at least in part by the dynamic nature of the tasks being performed, where the operation is changing over time. Procedures are needed that recognize and adapt to these changes.

In many cases, there are only two physical barriers to protect the workers, the pressure containment and PPE. Since many of the tasks involve opening the containment, adequate PPE is especially important. PPE requirements should be based on all possible release scenarios.

Recommendations for Industry and Academia

Process and Occupational Safety specialists should work together to understand and address small scale process incidents associated with common tasks. Some specific recommendations:

- Utilize this insight to better educate operations and maintenance personnel.
- Design process systems to minimize the need for hazardous tasks or reduce the severity of possible releases.
- Incorporate these findings in junior college process operator curriculum.
- Consider whether current procedures adequately cover the type of tasks most often associated with severe injuries and fatalities.

Further Investigation by the Authors

This study focused on CM and PR. Fatalities were addressed only for OSHA Region 6. The other 9 regions should be studied to confirm the trends and patterns and increase the confidence in the results. The severe injury data covered the 22 federal OSHA states. The scope for severe injuries and fatalities should be expanded to Oil and Gas and other industries.

The descriptions of the incidents in this study were extracted by the authors. To identify and interpret the entire dataset, machine learning techniques should be applied.

Develop systematic methods for identifying hazards for the most common tasks which were identified in this study.

Investigate the need for and characteristics of procedures that would prevent or mitigate the consequences of incidents from the common tasks.

References

https://www.osha.gov/data, This is the site for both data sources in this paper, "Fatality and Catastrophe Investigation Search" and "Severe Injury Reports".

Mannan etal, "Patterns and Trends in injuries due to chemicals based on OSHA occupational injuries and illness statistics", Journal of Hazardous Materials, 2008, p 349-356.

Chemical Safety and Hazard Investigation Board (CSB), Honeywell Incidents, Report No. 2003-13-I-LA, August 2005

Chemical Safety and Hazard Investigation Board (CSB), Toxic Chemical Release at the DuPont La Porte Chemical Facility, La Porte, Texas, Incident Date: November 15, 2014, No. 2015-01-I-TX

US Department of Labor, Phillips 66 Company's Houston Chemical Complex (HCC), Report to the President, April 1990.

618389 Incident Investigations to Enhance Safety Culture: Refinery Examples

Abstract:

The investigation of small incidents and near misses provides an opportunity for any company to change its safety culture at the grass roots level. This change is due in part to features unique to small incidents and near misses. The purpose of this paper is to outline the benefits of using small incidents and near misses to increase situational awareness and enhance safety culture. The paper will also include a refinery case study demonstrating how these small steps can make a big difference for process safety management.

Scope: The primary purpose of the investigation is to understand the underlying causes of the incident and to prevent it from happening over and over again. Events are not just human error or a result of a physical failure, each event is indicative of a fault in the system. The investigations are an important part of safety culture of company. They increase the awareness as well.

Methods, procedures, Process: The important elements of the event research are the timely research and report creation, the effective documentation of the findings and the sharing of the results. The method of the investigation is selected due to the potential risk of Incident/Near Miss/Safety Observation. In addition to the method, the assignment of team members is made by different levels of the organization depending on the level of potential risk assessment of the event.

Results, Observations, Conclusions: The current and potential evaluation of the impact of the incident to environment, human, reputation and to asset, the establishment of research teams, the appointment of root causes and actions, the automatic creation of the corrective and preventive actions or maintenance requests are made on specific investigation tab of the SAP (SAP EHSM) system. The publication of the accident research report are carried out by SAP EHSM as well.

In most of the applications Tier 1 and 2 Process Safety incidents are deeply investigated but sometimes Tier 3 and 4 incidents left behind in terms of investigation and actions completion speed. However, if you check all the Tier 3 and 4 incidents root causes comprehensively, you may see that they already call for the Tier 1 and 2 incidents. For example, you may find some corrosion reports as Tier 3 since it did not caused any Tier 1 or 2 incident. As you have investigation you will come up with some defaults due to crude oil quality or process technologies applied. Having such an investigation in the phase of Tier 3, you will be able to prevent such a big leakage. Moreover emphasizing on Tier 3 and 4 incidents investigation, gives the message that you are more proactive since you are not only dealing with the incidents only caused harm to people or asset but also potential risks. That will also help management to communicate to employees that, they are one-step ahead the incidents, which will lead employees to think potential risks in the workplace.

619133 Effective Implementation of Dhas – Trends One Year Past the NFPA Deadline

Abstract:

The deadline incorporated into NFPA 652 for existing facilities handling combustible dusts to complete a dust hazard analysis (DHA) has come and gone as of September 2020. How are companies doing with incorporating the findings of their DHAs into capital expense and operational expense budgets? Are authorities having jurisdiction (AHJs) even aware of the requirement? This presentation will review key findings from surveys of our clients and AHJs.

American Petroleum Institute

PROCESS SAFETY SITE ASSESSMENT PROGRAM

American Petroleum Institute
What Don't You Know about Your PSM Program?: How the Process Safety Site Assessment Program (PSSAP®) Can Help Sites Find and Address These Blind-Spots.
March 22, 2021

Often companies struggle to understand what process safety management areas need their focus. What specific topics am I neglecting? Which of my programs or procedures could use some help to be truly effective? How does my holistic program compare with my peers in the industry? The Process Safety Site Assessment Program (PSSAP®) is an industry-created, industry-led program that can help answer these questions.

The American Petroleum Institute (API) and the American Fuel and Petrochemical Manufacturers (AFPM) kicked-off the PSSAP program in 2012. Since then, API has administered over 130 of these assessments across the globe. These assessments cover the main focus areas of process safety management: Process Safety Leadership, Operating Practices, Mechanical Integrity (focused on fixed equipment), Safe Work Practices, Management of Change (MOC), Process Hazards Analysis (PHA), Facility Siting, Product Storage & Transfer, and Incident Learning. In addition to these traditional areas, the PSSAP program can also assess your Hydrofluoric Acid (HF) Alkylation/API RP-751 processes, as well as the Safeguards a site has in place to address risks, our newest offering.

But how can an assessment help my site? Won't I just receive a report? The PSSAP program is not a normal assessment. While our assessments are based on satisfying industry-created good practice protocols, an operator gets much more from our assessors than a report following their time on-site. API sends a full team of vetted, qualified assessors that are leaders in their field, with the average experience of one of our assessors being 40+ years. Our assessors spend time on site, and in addition to evaluating your programs for your blind spots, they also are constantly sharing the good-practices they have seen across the industry as part of their participation in this program. Not only does a site receive an assessment of their programs, they also can find out what the rest of the industry is doing to address these areas. Not only does a site get this through good-practice sharing from our assessors, but sites also get to see the benchmarking data from our previous assessments. With over 130 worldwide assessments, a site can statistically gauge how they compare with their peers. Sites also receive a detailed examination of the specific areas that could use more of their attention.

In addition to the different focus areas and our assessor pool, it is important to note that these assessments are scalable. Are you a small site? Is your operation too small for you to think such an assessment would be worthwhile? API created a shorter program for these sites, enabling them to access our industry-leading expert assessors so that all may benefit from their wealth of knowledge, continuing to drive refining, petrochemical and chemical operators toward our shared goal of zero process safety incidents.

We will explore our newest offerings in this forum, and will also show examples of what a site stands to gain from their participation in this industry-led and driven program.

619049 Bouncing Back from a Heart-Wrenching Process Safety Incident

Abstract:

During startup of Urea Fertilizer Plant after a brief shutdown, due to recurrence of a leakage, the field maintenance team was inspecting top head gasket of condenser on a 31-ft high platform when Carbamate solution splashed on the workers. They tried to escape via monkey ladder but fell down one after the other resulting in loss of life of one of the employees and hospitalization of other four.

Detailed incident investigation by a multi-disciplinary team highlighted number of gaps related to human elements, job handover/execution and crisis management. The incident occurred due to failure of multiple layers of protection including management of change, mechanical integrity, risk assessment, PTW, job supervision, fatigue element, over-occupancy, communication, training & development, emergency response, PPEs etc.

Interestingly, this happened in less than 3 years after successful compliance certification of Process Safety Management System but raw material unavailability and resulting financial crunch hugely jeopardized organization's process safety management system and safety culture.

The paper will cover this eye-opening case of huge PSM deterioration in very short time due to financial factors and how did the organization bounce back from that desperate situation. It will shed light on causes of incident, lessons learning and results achieved after revamp of the process safety management program. It will enable other participants to learn from story of our setbacks & revival, and implement similar cultural shift programs in their respective organizations.

SPRING21
+17TH GCPS
A Joint AIChE and CCPS Meeting

It Was a Dark and Stormy Night – Investigation of Acrylate Storage Tank Explosions

Kathleen Kas, Dow Chemical Company, Washington Crossing, PA
kkas@dow.com

Delmar "Trey" Morrison, Exponent Inc, Warrenville, IL
tmorrison@exponent.com

Prepared for Presentation at
American Institute of Chemical Engineers
2020 Spring Meeting and 17th Global Congress on Process Safety
Virtual
April 18 - 22, 2021

It Was a Dark and Stormy Night – Investigation of Acrylate Storage Tank Explosions

Kathleen Kas, Dow Chemical Company, Washington Crossing, PA
kkas@dow.com

Delmar "Trey" Morrison, Exponent Inc, Warrenville, IL
tmorrison@exponent.com

Keywords: Lightning, monomer, explosion, ignition

Abstract

It was in the early morning hours of May 16, 2012 when a thunderstorm rolled into the Bristol, Pennsylvania plant where The Dow Chemical Company produced acrylic polymers. As the weather turned stormy, lightning struck in the tank farm area containing inhibited ethyl acrylate, butyl acrylate, and methyl methacrylate. Within seconds, an ethyl acrylate tank exploded and was followed a few minutes later by a butyl acrylate tank explosion. The explosions and ensuing fire damaged the tank farm and led to a lengthy shutdown; fortunately, there were no injuries. This paper details the investigation into the complex sequence of events that led to the two explosions. The investigation team evaluated data gathered from eyewitnesses, security system cameras, weather services, and the process control system. The team also analyzed physical evidence including electrical bonding and grounding systems, deflagration arresters, and various tank components. Although the adjacent ethyl and butyl acrylate tanks were involved in the event, the investigation did find that flame isolation devices functioned as intended and prevented the propagation of flames to surrounding processes via the interconnected venting system. However, there was a puzzle left to unravel. The tank installations were determined to be in compliance with the industry lightning standards such as NFPA 780. If the tanks met industry standards, how did the explosion happen? While the exact mechanism could not be determined with certainty, the ignition of the atmosphere in the ethyl acrylate tank most likely occurred from a spark discharge inside the tank's headspace, likely between internal components at different electrical voltage potentials. This type of mechanism could invalidate the Faraday Cage assumption inherent in current industry guidance and may represent a hazard that is not currently well-recognized in industry. It should also be noted that, while lightning can't be prevented, the outcome of a lightning strike can also be mitigated by the use of safeguards such as inertion, flame arresters, diking, and foam systems.

1 Introduction

On May 16, 2012, at approximately 3:34 am, a lightning strike occurred at the Dow Bristol Plant in Bristol, Pennsylvania (Figure 1). Immediately after a lightning flash was observed, flames were observed on top of the Ethyl Acrylate Monomer (EA) tank about one second before it exploded. Liquid EA spilled into the dike shared with the Butyl Acrylate Monomer (BA) tank, and the BA tank subsequently exploded a few minutes later. The explosions and ensuing fire damaged the tank farm and led to a lengthy shutdown; fortunately, there were no injuries.

This paper details the investigation into the complex sequence of events that led to the two explosions. Key questions that the investigation team needed to answer included:

1. Was lightning the true initiating event?
2. Why did the contents of the EA tank ignite if the tank was grounded?
3. Why did the BA tank explode, since the headspace was not in the flammable region prior to the event?
4. What learnings can we take away from this incident?

This paper summarizes the event, the evidence collection and analyses, the testing of hypotheses, the conclusions from the event, and key learnings.

Figure 1 The dike fire and aftermath of the explosions.

2 Description of the Storage Tank System & Safeguards

The Dow Bristol Plant creates products for the Dow Coating Materials business. Acrylate-based coatings are produced at the facility. The facility receives, stores, and processes acrylate monomers including Ethyl Acrylate ("EA," CAS No. 140-88-5) and Butyl Acrylate ("BA," CAS No. 141-32-2).

2.1 Acrylate Properties

At room temperature, EA monomer is a colorless liquid with a pungent odor and a low odor threshold of approximately 1.2 ppb [1]. BA monomer is also a colorless liquid, with a fruity odor and an odor threshold of about 35 ppb.

The acrylate monomers are inhibited with the methyl ester of hydroquinone (MEHQ) to prevent unintended polymerization during storage. This inhibitor requires the presence of oxygen to retain activity. While systems exist to provide a mixed nitrogen/oxygen gas with a low enough oxygen level concentration to maintain both inhibitor activity and inertion, this type of mixed gas system has the potential to fail to an unsafe state. The limiting oxygen concentration for EA is 9.0 volume percent; thus, the control system to manage the atmospheric concentration would require a high level of reliability. The failure of the atmospheric control system can, and has, resulted in unintended polymerization in industry. In this installation, the risk of a runaway polymerization due to mixed gas failure was determined to be a greater risk than that of headspace ignition, so the tank was blanketed with air to maintain inhibitor efficacy.

The flash point of EA was below the ambient temperature of 19°C at the time of the incident, so the vapor headspace of the tank was flammable. The nearby BA tank's headspace, however, was not in the flammable region. The initial headspace vapor concentrations are shown on a vapor pressure curve for EA and BA in Figure 2.

Table 1. Physical properties of the acrylate monomers.

Property	**Ethyl Acrylate**	**Butyl Acrylate**
Flash Point	8°C	37°C
Lower Explosion Limit (LEL)	1.40%	1.30%
Upper Explosion Limit (UEL)	14.0%	9.90%
Maximum Experimental Safe Gap (MESG)	0.86 mm	0.88 mm
Odor Threshold	1.2 ppb	35 ppb

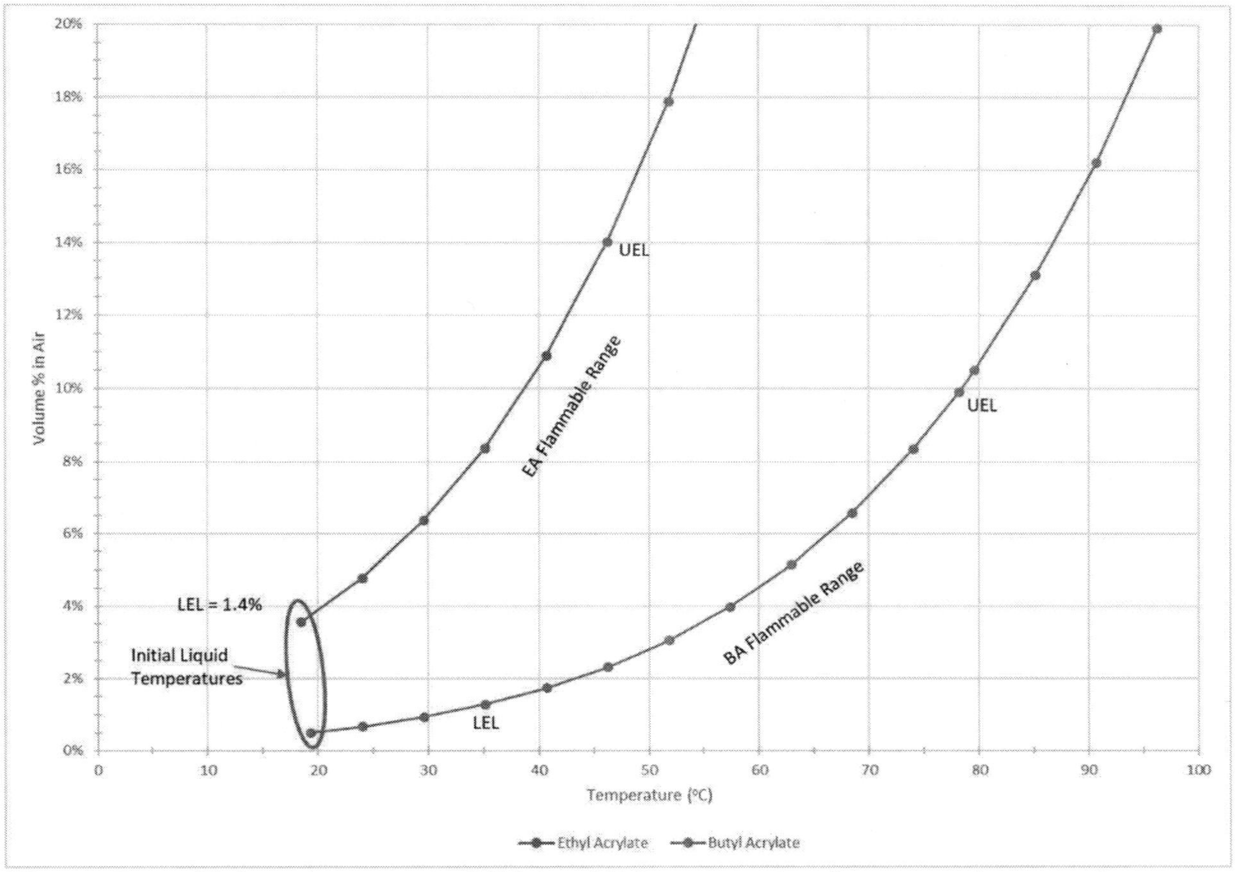

Figure 2 Volume percent of EA and BA vapors in air versus temperature.

2.2 Description of the Tank Farm

EA and BA monomers were stored in Tanks 120 and 119, respectively, in Tank Farm 30A.

Figure 3 is an aerial photograph of the area taken circa 2007. Liquid monomers were pumped into the EA and BA tanks from trucks and rail cars and transferred to the downstream manufacturing processes as needed. Truck and rail unloading stations are located around the perimeter of the tank farm dike walls. Figure 4 is a view of the tanks after the incident.

Figure 3 Aerial image of EA Tank (south) and BA Tank (north) in Tank Farm circa 2007.

Figure 4 View of tank farm after fire suppression efforts. Tanker truck in foreground was unloading MMA at the time of the incident.

The two atmospheric storage tanks involved in the incident contained inhibited BA liquid and inhibited EA liquid. The EA tank inventory was 673,482 lb and the BA tank held approximately 536,220 lb. At the time of the event, no transfers were occurring in or out of the EA or BA tanks. Truck unloading was occurring to a Methyl Methacrylate (MMA) tank in a separate tank farm area; however, the investigation concluded that MMA did not play a role in the event.

As shown in Figure 5, each storage tank was filled through an overhead dip tube and was equipped with a vapor header connection to the thermal oxidizer, an atmospheric conservation vent, and an atmospheric emergency pressure relief vent. The vapor header connection and atmospheric conservation vent were isolated from the tank using inline deflagration arrestors rated for this service.

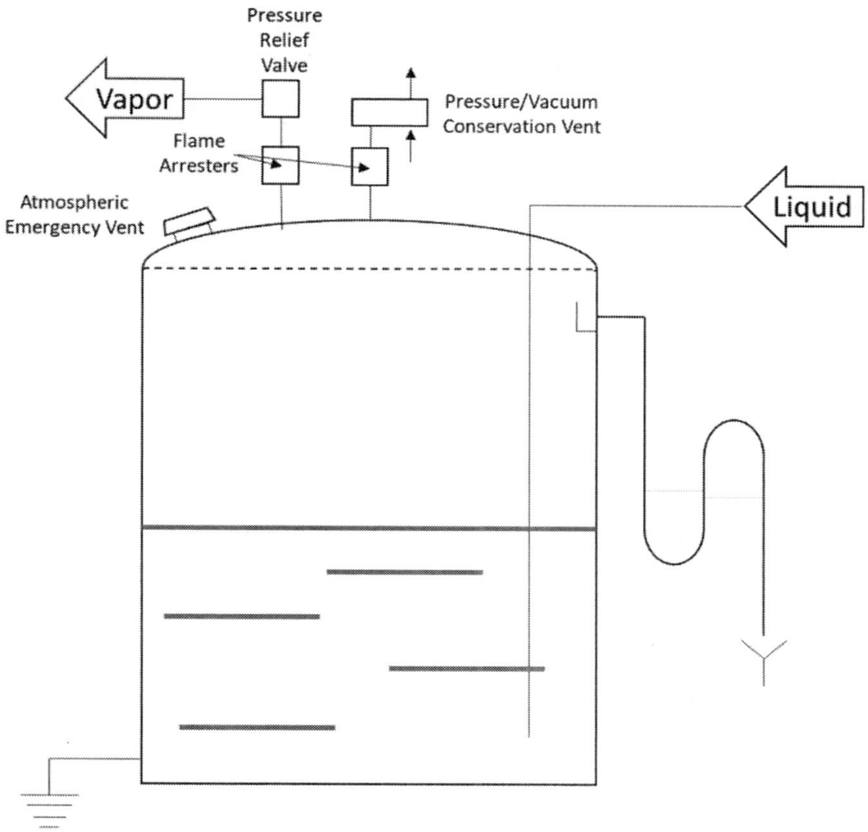

Figure 5 General tank appurtenance configuration and relationship.

2.3 Tank Construction

Tank 120 was in dedicated inhibited Ethyl Acrylate (EA) monomer service, and Tank 119 was in dedicated inhibited Butyl Acrylate (BA) monomer service. The carbon steel tanks were operated as atmospheric storage tanks and were originally constructed in the 1940s. The tanks were 30-feet in diameter by 29-feet tall with dome roofs. The dome roofs were originally constructed from petal sections with riveted overlap joints. At some point in the past, the rivets had been removed and the overlapped joints had been welded. The tanks were both 30 ft in diameter by 29 ft tall carbon steel cylindrical storage tanks with domes roofs. The steel thickness at the roofs and walls was approximately 0.3 inches. The tanks were equipped with a static grounding system and lightning protection consistent with guidelines in NFPA 780.

Table 1. Tank appurtenances.

Vent Device	Liquid Line	Instrumentation
Pressure/vacuum conservation vent to atmosphere	Liquid fill line	Capacitance level probe
Pressure relief valve to vent system (oxidizer)	Liquid withdrawal line	Liquid temperature probe
Pressure relief hatch to atmosphere	Overflow line with loop seal	Liquid temperature gauge
	Fire foam nozzle	

2.4 Tank Farm Grounding Installation

At the time of the event, National Fire Protection Association (NFPA) 780-2011, Standard for the Installation of Lightning Protection Systems, was a recognized industry standard. Section 7.2.2. of NFPA 780 advises: [2]

"Metallic structures that are electrically continuous; tightly sealed to prevent the escape of liquids, vapors, or gases; and of 3/16 in. (4.8mm) thickness or greater to withstand direct strikes... shall be considered to be inherently self-protecting."

American Petroleum Institute (API) Recommended Practice 2003, seventh edition (2008), also provided guidance. Section 5.4.1 stated: [3]

"Metallic tanks, equipment, and structures commonly found in the petroleum industry that are in direct contact with the ground (i.e. no nonconducting membranes) have proved to be sufficiently well grounded to provide for safe propagation to ground of lightning strokes. Supplemental grounding by means of driven ground rode neither decreases nor increases the probability of being struck, nor does it reduce the possibility of ignition of the contents."

The EA and BA tank installations at Bristol were found to be in compliance with these provisions. The vessels were steel tanks exceeding 3/16" thickness that were mounted on concrete pads and attached to grounding rods. Metallic piping and conduit were appropriately bonded to the tanks. There was an insulating membrane installed in the containment area; however, it did not isolate the tank and its associated pad from the connection to earth. Routine maintenance and testing were performed on the grounding system to maintain it in good working order. Therefore, no gap between industry guidance and the tank installations was identified.

2.5 *Deflagration Arresters*

Deflagration (flame) arresters were installed on the storage tanks to prevent the propagation of deflagrations from an external source into a tank storing combustible or flammable material (Figure 6). Flame arresters typically contain small, high surface area passages through which momentum and heat of a flame front passing through a flammable vapor/air mixture in piping are quenched. Flame arresters are sized according to the chemicals involved. NFPA 497 reports maximum experimental safe gap (MESG) values between the plates of 0.86 mm for ethyl acrylate (inhibited) and 0.88 mm for n-butyl acrylate (inhibited).[4] These gap sizes correspond to Class I Group D materials, or those materials with an MESG greater than 0.8 mm.[5]

Figure 6 Example pressure/vacuum relief device (VPRV) mounted above a deflagration arrester.

3 Investigation

The investigation was initiated once the fire was extinguished and the scene was rendered safe. The focus of the investigation was to gather data to understand the sequence of events, identify the ignition mechanism for the EA tank, and determine the means by which the BA tank also ignited.

3.1 Understanding the Sequence of Events

3.1.1 Eyewitness accounts

Two witnesses were located in the area of the tank farm at the time of the lightning strike and explosion, with positions shown in Figure 7. None of the witnesses saw the location of the lightning strike. One witness (Witness #1) was sitting in a truck by the MMA unloading station completing paperwork. He saw the flash and heard the lightning strike, which was shortly followed by the EA tank explosion. A second witness (Witness #2) was located in a nearby building, sitting at a desk on the east side. He looked up immediately after the strike and noted flames on top of the EA tank, which preceded the explosion.

Figure 7 Witness locations to the lightning strike and ensuing events.

The lightning discharge itself was not captured by surveillance video; however, a flash associated with the lightning strike was recorded by a camera pointing to the northwest.

3.1.2 Evidence collection

Immediately after the emergency response activities were completed, the scene was secured. Photographs were taken of the tank farm and surrounding areas. A chain-of-custody process was established for evidence. Debris from the explosions was located, tagged, photographed, logged, bagged, and retained in a locked area. Security camera footage from all site cameras was secured. The Distributed Control System (DCS) data was retrieved and archived so that it would not be overwritten.

3.1.3 Verification of lightning strike

At the time identified in security video and supported by witness testimony, a lightning strike to ground was recorded by two lightning verification services: StrikeNet and TOA. Both organizations recorded a strike to ground within the Bristol site with a magnitude between -36 kA and -38kA. Although the sensors used by the two organizations placed the lightning strike within the Bristol site at slightly different locations, the tank farm fell within the confidence zones for both sets of data.

The equipment, debris, and surrounding areas were inspected after the event to determine if the locations of one or more lightning attachment points could be located; however, no burn marks or other indications of a lightning strike were evident on the steel piping or equipment in the tank farm or connected to equipment in the tank farm. Attempts to detect erratic magnetic fields on steel equipment were also inconclusive. However, based on the eyewitness testimony, security camera footage, and lightning strike data, it was concluded that a lightning strike had occurred and initiated the event.

3.2 Evaluation of Physical Evidence

3.2.1 Review of tank damage

About a second after the lightning strike, the EA tank exploded (Figure 8). The explosion displaced the dome roof, separated piping systems, and damaged the lower tank shell. The explosion yielded an immediate fireball followed by a tank fire and pool fire in the surrounding dike area from burning EA monomer liquid. A post-incident photograph of the EA tank is shown in Figure 9. Flames were observed at dome level immediately after the lightning strike by an employee in a building about 100 feet southwest of the tank. The roof of the EA tank fractured into multiple pieces, and a portion was projected outward and hit a nearby trailer. A large section of the roof, including the deflagration arrester and vacuum/pressure relief assembly, fell into the tank. The northwestern and western portions of the upper-most segment of the sidewall of the tank, as well as the external ladder of the tank, were bent inward. The flat tank bottom bowed due to the internal pressure prior to rupture of the tank roof.

Figure 8 **First frame of surveillance video showing the EA tank explosion. This frame is time-stamped 1 second after the frame capturing the lightning strike.**

Figure 9 **Ethyl acrylate tank after the incident. The upper portion of the tank was curled into the tank from when the roof fell back into the tank after the explosion. Oxidation and soot deposition are evident on the outside of the tank wall.**

Approximately two minutes after the EA tank exploded, the BA tank also ruptured (Figure 10). The roof was found on a nearby loading rack with apparent soot deposition but with paint intact. The dome was curved inward and several of the radial seams of the roof were separated. A photograph of the dome appears in Figure 11.

Figure 10 BA tank from the west.

Figure 11 BA tank dome, which landed on a nearby loading rack.

3.2.2 Deflagration Arresters

The EA tank conservation vent device was found to be damaged but still attached to the tank's dome. The overpressure relief component of the conservation vent was separated from the rest of the assembly. The pressure relief valve pallet had been displaced upward, and the weather hood studs had been pulled out of the assembly flange. There was evidence that the guide cap impacted on the top weight, and the guide cap rivet was pulled out. The fluorinated ethylene propylene (FEP) polymer seal on the pallet appeared heat-affected.

As shown in Figure 12, the EA tank's conservation vent flame arrester internal cartridge was damaged mechanically – the plates were displaced. Examination of the plates in the cartridge revealed that the tank headspace side of the cartridge displayed a circular pattern of erosion indicative of an unknown duration exposure to flames venting from the tank into the flame arrester. The atmospheric side of the cartridge did not exhibit a similar heat-affected damage pattern. The presence of erosion on the tank side of the cartridge but not on the opposing side indicated that flame travel through the arrester was directed from the inside of the tank to the outside. This confirmed that the flame arrester operated as intended and did not allow flames to vent from the tank. This also supports a conclusion that the flame arrester cartridge would have prevented an external fire from passing into the tank headspace.

Figure 12 EA tank conservation vent deflagration arrester cartridge. Tank side on the left and the outside is on the right. Flame impingement pattern highlighted on inside of cartridge.

As shown in Figure 13, the BA tank conservation vent flame arrester showed similar flame impingement patterns from the inside of the tank as the EA tank.

Figure 13 BA conservation vent deflagration arrester cartridge. Tank side on left and outside on right. Flame impingement pattern highlighted on inside of cartridge.

The BA tank's vent line inline deflagration arrester was ejected from the tank and later found inside the tank after being subjected to the pool fire inside the tank. The deflagration arrester was too heavily damaged to determine if it was attacked from an internal or an external fire. However, the EA tank vent line flame arrester was intact enough to conclude that the EA tank fire was unlikely to communicate with the BA tank through the vent header. The EA tank vent line was equipped with a 3-inch inline flame arrester, shown in Figure 14. The flame arrester was located between the tank and the vent header, which went to the thermal oxidizer. The flame arrester assembly was fragmented and pieces were projected in multiple directions away from the tank. The impact caused collapse and deformation of the corrugated aluminum plates in the internal cartridge based upon the lack of fire/explosion damage to the edges of the plates. The plate array also displayed a circular pattern of erosion indicative of an unknown duration of exposure to flames venting from the tank into the flame arrester. The atmospheric side of the cartridge did not exhibit a similar heat-affected damage pattern. The plate surfaces were also free of heat-affected damage patterns. The presence of erosion on the tank side of the cartridge but not on the opposing side indicates that flame travel through the arrester was directed from the inside of the tank to the outside. As shown in Figure **15**, the clean plate surfaces indicate that the flame penetrated a minimal depth into the cartridge. This indicates that the flame arrester operated as intended and did not allow flames to vent from the tank. This also indicates that an external fire, for example due to the ignition of a release of flammable vapor through the shared vent line, did not pass into the tank through the arrester.

Figure 14 EA vent line inline deflagration arrester cartridge. Tank side on left and outside on right. Flame impingement pattern highlighted on inside of cartridge.

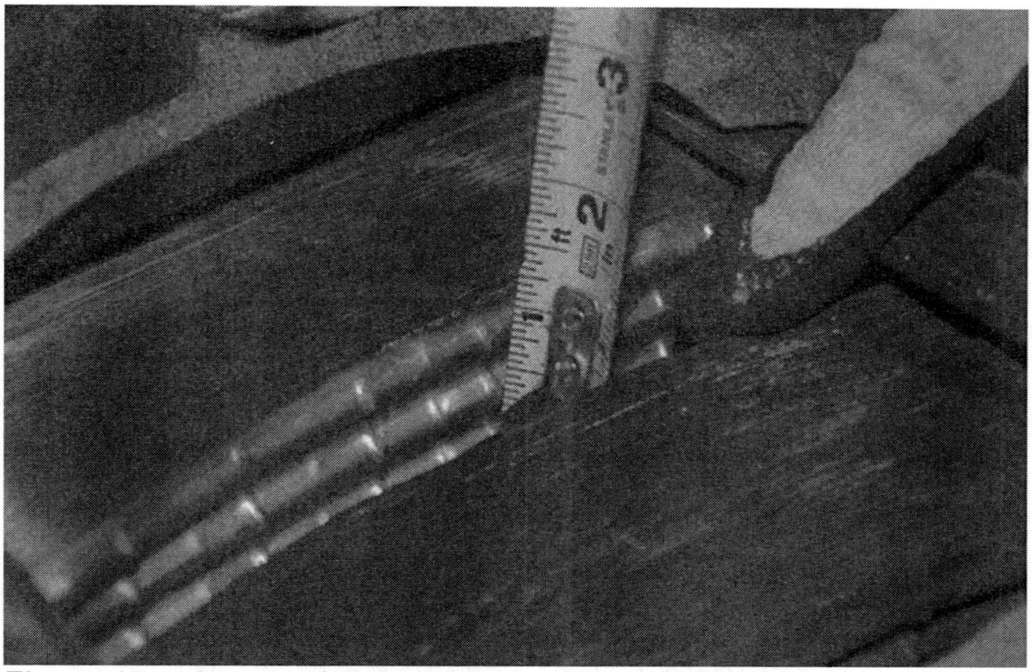

Figure 15 **Image of clean plate surface inside cartridge indicating that flame penetration depth into the cartridge was minimal.**

3.2.3 Location of EA tank headspace ignition

After the lightning strike, there was a time delay before the EA tank exploded. Based on the witness statements, security camera footage, and recorded process data, the following sequence of events was identified for the explosion. Table 3 provides the sequence of events.

Table 3. Sequence of events.

3:47:08	Lightning strike
	Flames visible on roof of EA tank
3:47:09	Explosion of EA tank
	EA tank fire and pool fire in dike
	BA tank explosion

The duration of the delay was approximately 1 to 1.2 seconds. During this interval, Witness #2 reported seeing flames on top of the EA tank but could not be more specific about the nature of those flames. Possible explanations include ignition of a pre-existing vapor release from one of the roof appurtenances or a crack in the roof, burning of a combustible solid material such as a plastic component, or venting of flames from the emergency vent. There are many parameters that may have affected this timing, including the flame speed in EA vapor/air mixtures and the effects of pressure relief devices. The approximate method presented in the SFPE Handbook was used to estimate the pressure rise with time inside the EA tank head space by assuming propane combustion properties for the EA vapors (Equation 1).[6] Given these calculations, the estimated burst pressure for the vessel of approximately 2 psig would be reached in about 1 second, as shown in the graph in Figure 16. Thus, the 1 to 1.2 second delay between the lightning strike (ignition) and explosion is reasonable and consistent with the delay expected for an internal deflagration to cause the tank to rupture.

Equation 1:

$$\frac{P-P_0}{P_0}=\left(\frac{1}{\gamma}+\frac{P_0}{P_m-P_0}\right)^2\left(\frac{P_m-P_0}{P_0}\right)^3\left(\frac{\chi S_u t}{a}\right)^3 \ll 1 \quad (7)$$

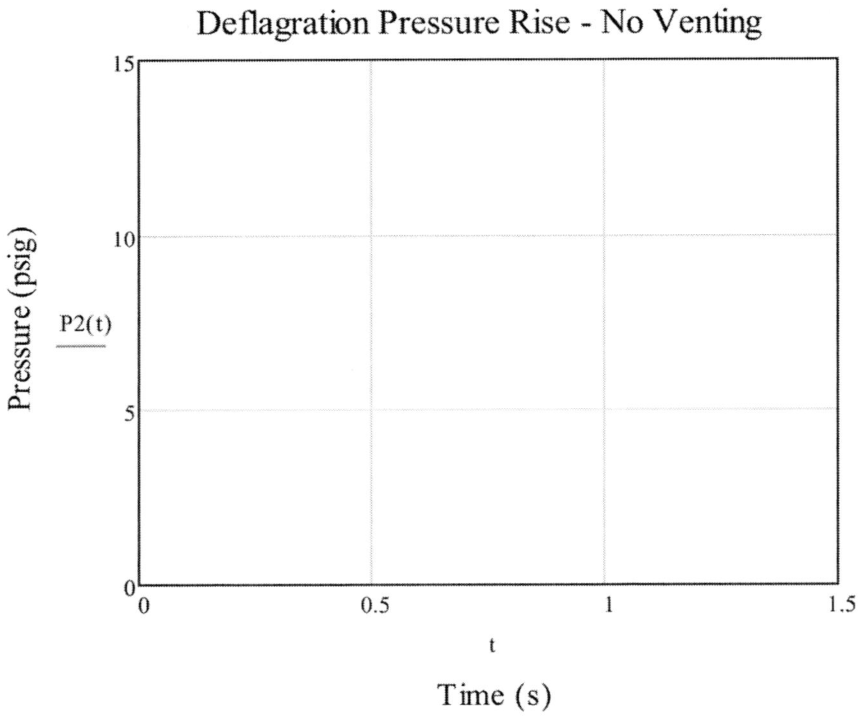

Figure 16 Example pressure rise curve versus time for propane combustion properties assigned to EA tank head space.

Based on this calculation and logical deduction, the point of ignition inside the headspace was likely near the 12-inch emergency vent such that flames would be issuing from the vent for a period of time prior to tank rupture pressure being reached. The emergency vent would be anticipated to open quickly after an internal pressure rise begins, which would allow flames to vent. If the internal deflagration had originated away from this area, the witness would have only observed flames close to or coincident with the explosion.

3.3 Testing Hypotheses

For a fire to occur, it is necessary to have a fuel, an oxidant, and an ignition source. In this case, the flashpoint of EA was 8°C – well below the 19°C temperature of the liquid. The tank contents were stored under an air atmosphere. Since fuel and oxygen were both present in the EA tank, the investigation focused on determining potential ignition mechanisms for the EA tank.

While the headspace of the EA tank was flammable, the flash point of BA is 37°C – above the ambient temperature at the time of the event. As a result, hypotheses were developed to determine not only the ignition source for the BA tank but also how the tank temperature increased to the point where the contents became flammable.

3.4 Hypotheses: Ignition of the EA Tank

Three potential sources of the EA tank ignition were evaluated:

1. Ignition source entered the EA tank via the vent system
2. Ignition source entered the EA tank while it was breathing/venting to atmosphere
3. Ignition source internal to the tank.

3.4.1 <u>EA Hypothesis 1: Ignition source entered the tank via the vent system</u>

The vent header could possibly have contained a flammable mixture due to the MMA truck unloading activities that would displace the MMA tank flammable headspace into the vent header. However, ignition in the vent header system outside of the tanks was eliminated as a potential cause based upon several observations. The fire pattern present on the EA tank vent header deflagration arrester was located on the inside (i.e., tank side) of the deflagration arrester, indicating that flames were venting into the arrester cartridge for some period of time prior to the explosion. Additional evidence that discounted external ignition in the vent system included the lack of soot in examined piping, the lack of overpressure damage in piping and connected tank components, and the lack of indication of failure of the PRV isolating the tank from the vent header.

3.4.2 <u>EA Hypothesis 2: External ignition source while EA tank was breathing/venting</u>

Although the EA tank was not directly associated with the filling activities ongoing at the time of the incident, the possibility for flammable EA vapors to leak from the tank was evaluated as a potential path to internal ignition. Since the EA tank was anticipated to act as a Faraday-like cage and prevent penetration of an electrical discharge, the possibility of an external discharge igniting a vapor leak was considered. Four primary leak locations were identified and evaluated:

1. Release from a crack in the shell or dome roof
2. Release from an incompletely filled overflow line loop seal

3. Release from EA conservation vent
4. Release from EA 12-inch emergency vent

EA vapors have a very low detection threshold (0.5 ppb), and from the investigation team's experience, a leak in the tank would have been detected during normal operator inspections if it was near ground level. A small leak at roof level could not be as easily detected. A recent tank inspection did not detect cracks in the tank shell or vapor leaks in the tank. If such a small crack existed, it would be unlikely to be larger than the minimum experimental safe gap (MESG) of 0.86 mm; thus, it was unlikely that ignition could propagate back into the tank through such a crack. Therefore, a leak from a crack in the shell or dome roof was refuted as a potential ignition pathway.

The loop seal fluid had been replaced in 2010, and a leak past the seal fluid would be anticipated to be detected by field operators since the opening is near ground level adjacent to the tank. Thus, a leak from the loop seal was refuted as a potential source for vapors.

An ignition pathway through the EA conservation vent was also considered. However, the deflagration arrester did not exhibit signs of impingement by an external flame. Additionally, the deflagration arrester appeared to extinguish the internal flames as intended before they could pass through the cartridge. The deflagration arrester cartridge was packed with parallel plates and suitable for suppressing flames in either flow direction. Thus, an external ignition at the VPRV that flashed back into the tank to ignite the tank headspace vapors was refuted as a possible cause.

The larger 12-inch diameter emergency vent was not equipped with a deflagration arrester. The pallet assembly had a rigid FEP film diaphragm seal, which was an electrical insulator. The device was destroyed in the incident, but most major components were recovered. From the recovered components, no evidence was found of electrical bonding (e.g., bonding strap) between the pallet and the frame or tank. The listed set point for the emergency vent was the same (1.5 psig) as that for the conservation vent; thus, the tank could have breathed out through either vent. An ignition at this emergency pressure vent assembly could not be refuted during the investigation.

3.4.3 EA Hypothesis 3: Ignition source generated from an internal spark

As indicated in NFPA 780-2011 and API 2003 (2008), metallic tanks that are electrically continuous, sealed, and in direct contact with the ground are considered to be protected from lightning. This is based on the assumption that the steel tank system acts like a Faraday cage. A Faraday cage is an electrically conductive enclosure that protects the interior of the enclosure from the effects of external electrical fields. Electrical charges are distributed and flow to ground, cancelling the field's effect on the interior.

However, the protection offered by the Faraday cage can be negated by the presence of an unbonded internal component. As API 2003 (2008) pointed out:

"Metal tanks with fixed metal roofs and horizontal metal tanks that are maintained in good condition are generally protected from damage by direct-stroke lightning and ignition of its contents, if all metal components are in electrical contact (i.e. bonded)."

It was therefore possible that direct internal ignition of the EA vapors inside the tank could have occurred due to a spark discharge across a gap between two internal tank components that were not electrically bonded with each other. A detailed analysis of the tank instrumentation, equipment, and appurtenances yielded a list of several potential ignition locations. Piping penetrations into the headspace were either refuted as non-credible or assigned a low likelihood given the nature of mechanical contact between piping and the tank shell.

Conservation and emergency relief vents

The conservation vent and the emergency relief vent both utilized weighted floating pallets with FEP diaphragm seals. The pallets were not electrically bonded to the device bodies or the tank shell. Given the discontinuous electrical nature, it is possible that a spark discharge occurred inside either assembly. However, the conservation vent was separated from the tank headspace by the deflagration arrester. Thus, the conservation vent was refuted as a potential ignition source.

The pallet manway was assumed to behave as a capacitor in the event of a lightning strike, and calculations were performed to estimate the energy of a spark that could have occurred between the pallet and the tank. The calculations indicated that the energy generated would only reach about half the minimum ignition energy (MIE) of ethyl acrylate. However, due to uncertainty in the estimation, the emergency relief vent could not be refuted based on the available information.

High level capacitance probe

The tank was also equipped with a high-level capacitance probe. It was determined that the electrical ground lead inside the sensor housing was not connected to an earth ground, which may have affected the device's ability to dissipate electrical charge. It was noted, however, that the circuit breaker remained active and did not trip during the event. An examination of the level switch showed that the fuse was intact and there were no signs of overvoltage/overcurrent. However, it appeared that the diodes between probe and shield inputs may have been shorted, which could have occurred if the arc were limited to a few microseconds of exposure. Although evidence was not found to indicate that a spark actually occurred between the high-level probe and the tank, the potential for a spark discharge from the probe to a surface inside the tank could not be entirely refuted.

Surface charge

The team also considered the possibility of a liquid surface charge creating a spark discharge. However, this was deemed unlikely based on the conductivity of EA.

3.4.4 Plausible hypotheses: EA tank ignition

Based on this evaluation, the most likely ignition pathway was an electrical discharge from one of two electrically isolated equipment components internal to the EA tank – the high-level probe or the pallet manway.

3.5 Hypotheses: Ignition of the BA Tank

Although the headspace in the EA tank was flammable, the BA liquid was below its flashpoint and was not in a flammable region. Thus, only one side of the fire triangle was present. For ignition to have occurred within the BA tank approximately two minutes after the lightning strike, a source of ignition energy must have made its way into the tank head space, and the tank head space must have had a flammable concentration of vapors. Hypotheses were therefore generated to explain both the generation of an ignitable mixture of BA in air and the source of ignition.

3.5.1 Creating a flammable atmosphere in the BA tank

The vapor space of the BA tank was not flammable under normal operating conditions. The measured temperature of 19.4°C prior to the incident was well below the flash point of 37°C reported in the Safety Data Sheet. However, there must have been a flammable composition in the vapor space of the BA tank at the time when the tank exploded approximately two minutes after the EA tank explosion. The BA tank was in close proximity to the burning EA tank, and the two tanks shared a common dike area. Heat output from the EA fire in the dike, and possibly heat from the EA tank itself, could have raised the temperature of the BA tank, enriching the vapor space in BA vapors. The temperature probe in the BA tank was located in the lower part of the tank and ceased its output to the control system a little more than a minute after the EA tank erupted. Prior to its failure, the thermocouple had registered no appreciable change in temperature. However, given the size of the tank and proximity to the fire, it is very likely that a portion of the tank contents was raised significantly in temperature from the radiant heat of the EA tank fire and potentially direct flame impingement, possibly creating boiling at the shell interface.

One potential outcome of boiling inside a tank due to pool fire exposure is a Boiling Liquid Expanding Vapor Explosion (BLEVE). This type of explosion is possible if the tank is a rated pressure vessel and pressure relief systems are unable to maintain a safe pressure during fire exposure. Since the BA tank was an atmospheric storage tank and the shell did not display bulges, a BLEVE was refuted as a potential cause of explosion.

Three hypothetical paths to ignition of the BA tank vapor space were identified: ignition in the vent pipe which entered the tank, ignition inside of the tank, and ignition outside of the tank which propagated inside.

3.5.2 BA Hypothesis 1: Communication through the vent header

At the time of explosion of the BA tank, the vent header system had been compromised at the EA tank. Despite this damage, ignition in the vent header system outside of the BA tank was eliminated as a potential cause based upon observations of the deflagration arresters. Although the vent line deflagration arrester was heavily damaged and conclusions could not be drawn, the counterpart deflagration arrester beneath the conservation vent displayed internal fire patterns, not external patterns. This model of flame arrester appeared to successfully extinguish flames in EA and BA vapor streams based upon analysis of the surviving components.

3.5.3 BA Hypothesis 2: Ignition through the conservation vent

Because of the fire damage to the conservation vent flame arrester on the BA tank, no direct evidence of functionality could be determined. However, there were no apparent gaps in the plates in the cartridge. Given its rating for this service, it is unlikely that BA vapors venting through the conservation vent could flash back through the device into the tank headspace.

3.5.4 BA Hypothesis 3: Hot surface ignition by the exterior pool fire

Ignition inside of the tank could have occurred due to hot surface ignition within the vapor space. The low BA auto-ignition temperature of 292°C could possibly have been achieved on the inner tank shell surface from radiant heat or direct flame impingement at or near the liquid level where BA vapors would be anticipated to be within the flammable range. The BA tank shell was heavily heat-affected by the fire, and all paint had been heavily damaged. The tank roof was still painted, but the side facing the EA tank was soot covered indicating fire exposure. Thus, this pathway of hot surface ignition inside the tank head space could not be eliminated as a hypothetical cause.

3.5.5 BA Hypothesis 4: Direct ignition by the surrounding fire

Given exposure of the tank to the surrounding fire and the evidence of an internal deflagration, it is apparent that the liquid inside the BA tank was generating vapors. The BA tank was equipped with both a conservation vent and an emergency pressure vent. Since the conservation vent and the 12-inch emergency pressure vent pallet manway were set at the same relief pressure; it is likely that both vents were relieving BA vapors during the incident. Under this condition, the nearby EA tank and EA liquid pool fires could have directly ignited the vented vapors if flames contacted the vapors. The resulting ignition at the pallet manway would have allowed a flash back into the tank head space. This potential ignition pathway was a likely cause for the BA tank explosion.

3.5.6 Plausible hypotheses: Ignition of the BA tank

Although autoignition of the BA as a result of direct flame impingement could not be ruled out, the more likely ignition pathway was ignition of BA vapors venting from the emergency pressure vent pallet manway and flashing back into the tank.

4 Analysis and Conclusions

While the location of the lighting strike could not be precisely determined, the investigation concluded that the lightning strike caused ignition of the flammable head space vapors in the EA tank. The resulting internal deflagration caused the tank's internal pressure to quickly rise to a level causing the tank dome roof to rupture. The subsequent EA liquid pool fire in the containment dike and in the ruptured EA tank caused the BA tank, which did not normally contain a flammable vapor head space, to raise the head space concentration into the flammable range such that internal deflagration could occur.

The most likely cause of ignition of the EA vapors was determined to be an internal spark discharge from a non-electrically bonded component. The two most likely candidate components were the high-level capacitance probe and the 12-inch diameter emergency pressure vent pallet manway. The most likely cause of ignition of the BA vapors was determined to be either hot surface ignition by the internal shell above the liquid level or direct ignition of vapors vented from the 12-inch diameter emergency pressure vent pallet manway by the surrounding fire.

Following the event, Dow created a corporate learning experience report that was shared throughout the organization. It described the event, detailed the root and contributing causes, and discussed key learnings. A number of follow-up actions were also created. These included:

- Review operations policies and procedures concerning suspending manual loading and unloading of flammable materials when there is nearby lightning. Lightning guidance from the National Weather Service [7] recommends stopping outdoor activities from the time when lightning is first seen/heard until 30 minutes after the last clap of thunder.
- Review flame arrester maintenance procedures to ensure that they include plate spacing observations and proper methods for cleaning. Although the flame arresters were found to have worked as designed in this event, it was a good opportunity for the organization to reaffirm that all flames arresters are being properly serviced.
- Review maintenance procedures for inspection and testing of tank grounds and lightning apparatus to confirm that all are being maintained and work as designed.
- For storage tanks that can have a flammable headspace, complete an assessment of the tank, which included:
    - Ensuring that hatches are closed when not in use
    - Ensuring that roofs are in good condition, with no non-conductive patches
    - Evaluating the installation for the appropriate use of vacuum vent valves, flame arresters, gas padding, and other means to prevent flame propagation into the tank.
    - Assessing the equipment for tank penetrations that are not bonded to the tank shell or are not electrically continuous throughout their construction. These could include dip pipes, float level indicators, etc.

Action plans were developed based on the opportunities that were identified from this review.

5 Lessons Learned

Key takeaway messages from this investigation include:

- Lighting strikes are powerful natural events that can't be prevented. It has been estimated that lightning strikes would be expected to increase by about 12 percent per degree Celsius of warming [8]. It is critical to design our equipment and facilities to withstand extreme weather events like lightning.

- Following industry standards on lighting will not prevent all ignition events.

  As stated in API Recommended Practice 2003 (2008), Protection Against Ignitions Arising Out of Static, Lightning, and Stray Currents Section 5.1, it should be noted that: *"Even when all known precautions are employed, prevention or safe dissipation of direct-stroke lightning cannot be absolutely assured. In the case of indirect lightning currents, incendive sparks may occur in some segments of a system that use the best-known precautionary methods and devices."*

- Current standards may not protect against the potential hazard of unbonded components.

  The Faraday cage assumption inherent in the assumption of grounding in the event of a lightning strike assumes that there are no unbonded components. Should that assumption be invalidated, a system that otherwise complies with industry lightning guidance and best practices may not actually be protective. An evaluation of uninerted tanks in flammable service is recommended to identify and remediate any unbonded components.

- To mitigate the risk of lightning, it is prudent to employ good industry practices against ignition, such as flame arresters, closed operation and venting, and grounding. Also, when installing new tanks, consider the proximity to existing tanks. The BA tank likely would not have ignited had it not been in close proximity to the EA tank and had not shared a common dike.

- Consider inerting flammable tanks in situations where the risk of ignition outweighs the risk of failure of the inerting or mixed gas system.

6 References

[1] ACRYLATE ESTERS A SUMMARY OF SAFETY AND HANDLING 3RD EDITION Acrylate Esters, A Summary of Safety and Handling, 3rd edition, 2002. Available at ICSHAM Acrylic Esters Safe Handling Guide (dow.com).

[2] National Fire Protection Association (NFPA) 780, Standard for the Installation of Lightning Protection Systems, 2020 edition.

[3] American Petroleum Institute (API), Recommended Practice 2003, seventh edition (2008).

[4] NFPA 497, Recommended Practices for the Classification of Flammable Liquids, Gases, or Vapors and of Hazardous (Classified) Locations for Electrical Installations in Chemical Process Areas, 2012 edition.

[5] Grossel, S.S., *Deflagration and Detonation Flame Arrestors*, CCPS (2002).

[6] Society of Fire Protection Engineers (SFPE), Handbook of Fire Protection Engineering, 3rd Edition, section "Closed Vessel Deflagrations," pages 3-406 to 3-410 (2002).

[7] National Weather Service, https://www.weather.gov/safety/lightning-safety-overview, accessed March 14, 2021.

[8] Scientific American, https://www.scientificamerican.com/article/lightning-may-increase-with-global-warming/, accessed March 14, 2021.

618785 Dam(n) Safety Lessons: Risk Based Process Safety Principles Applied to a 2020 Dam Failure

Abstract:

On May 19, 2020, over 10,000 Mid-Michigan residents were ordered to evacuate their homes during a pandemic because of two impending dam failures. Swift emergency response coordination enabled people to evacuate in time and no one lost their lives. Unfortunately, nearly every business in the village of Sanford, MI was destroyed; more than 2,000 homes were damaged or condemned; roadways and bridges were swept away; and two lakes disappeared within hours, resulting in permanent disruption to the local ecosystem.

This disaster could have been avoided: The dam system had experienced three near misses in its 95-year history prior to failure, and deficiencies were discussed between the dam owners, government, and community for over 11 years. In addition, dam operators financially struggled in a significantly changing business model for small-scale hydroelectric production. Interestingly, this experience is not unique to dams and infrastructure; it can be applied to companies that manufacture, use, transport and store hazardous materials.

Process safety is the study of applying management systems to prevent rare, catastrophic events such as fires, explosions and toxic releases. The probability of these catastrophic events is rare, but when they do occur, they can destroy the lives of workers, the community, and the reputation of the organization.

We can apply the elements of Risk Based Process Safety (RBPS) to the 2020 Mid-Michigan dam failures as a case study to demonstrate where systems failed and understand how these risks could be averted in the future. This case study will show the impact of using RBPS to prevent rare, catastrophic events at many types of manufacturing and warehousing facilities.

This paper will walk through the four pillars of the CCPS Risk Based Process Safety Model, using publicly available data from the 2020 Michigan dam failures as a case study to illustrate how impactful the RBPS pillars can be for many non-chemical industry segments:

1. **Commit to Process Safety**, including the importance of leaders giving authority to competent personnel to escalate concerns so that concerns can be identified and resolved swiftly, and to perform due-diligence audits prior to acquiring a high-risk operation.
2. **Understand Hazards and Risks**, including the success story where Emergency Responders used knowledge of the projected worst-case flood analysis to quickly evacuate at-risk member of the community.
3. **Manage Risk**, including maintaining equipment to meet Recognized and Generally Accepted Good Engineering Practices (RAGAGEP), and to swiftly resolve deficiencies once discovered.
4. **Learn from Experience**, including ensuring someone with authority reviews facility incident data; Industry incident data; and third-party audit inspection results to shift resources to fix a systems issue before disaster strikes.

SPRING21
+17TH GCPS
A Joint AIChE and CCPS Meeting

Which eye are you looking through? The impact of cognitive bias on process safety

Trish Kerin
IChemE Safety Centre
Melbourne, Australia
tkerin@icheme.org

Prepared for Presentation at
American Institute of Chemical Engineers
2021 Spring Meeting and 17th Global Congress on Process Safety
Virtual
April 18 - 22, 2021

Which eye are you looking through? The impact of cognitive bias on process safety

Trish Kerin
IChemE Safety Centre
Melbourne, Australia
tkerin@icheme.org

Keywords: bias, culture, leadership, lessons learned

Abstract

Bias surrounds us every day. Everything we look at has a lens of our experience coloring our perception, and everyone has their own lens. This impacts how we share and receive information. So why is this important for process safety? Cognitive biases cover a range of aspects, each of which can have an impact on process safety, from what we remember, to coping with too much information, to a lack of context and meaning or a need to act with great haste. There are many identified specific cognitive biases, not every one of them impacts on process safety, but many can and have. This paper will review several process safety type incidents that had cognitive bias as a factor and discuss ideas on how to manage the bias, so we do not fall into its trap.

1 Cognitive bias

Cognitive bias is defined as a systematic deviation from norms or rational judgement, a way of creating a subjective based reality based on an individual's perceptions [1]. While on the surface this sounds like a pitfall in human decision making, our cognitive biases allow us to make quick decisions. This can be advantageous when time is of the essence, provided the bias does not lead us in the wrong direction. Making decisions based on biases, or heuristics to estimate outcomes is a faster way to make decision than rational decision making, and it also fits in with the idea that we have two mind systems, system 1 and system 2. "System 1 operates automatically and quickly, with little or no effort and no sense of voluntary control. System 2 allocates attention to the effortful mental activities that demand it, including complex computations." [2, p. 20]. The problem with system 1 is that it is sometimes wrong, and unless we use system 2 to override the wrong answer, we can end up with the wrong decision. Take for example Figure 1. Quickly glance at it an decide which figure is the largest. Your system 1 would have said the one on the right.

Figure 1 – Which figure is larger? [2, p. 100]

If you were to engage your system 2, and take a ruler and measure the figures you would see they are all the same size. System 1 was misled by a perspective perception. The challenge for managing bias is to engage system 2 prior to making critical decisions.

Another concept to describe this System 1 thinking is called "thin slicing". This "refers to the ability of our unconscious to find patterns in situations and behavior based on very narrow slices of experience." [3, p. 23] It can be summarized as that initial gut decision we may feel in an instant.

1.1 Most common biases

There are 188 identified different cognitive biases, though they can be grouped into the 4 categories of:

- What should we remember?
- We need to act fast
- Too much information
- Not enough meaning [4]

Common biases that can have an impact on safety are:

- Anchoring
- Confirmation bias
- Framing
- Hindsight bias
- Status quo bias
- Sunk-cost bias [5]

In addition to these six, Halo bias can lead to giving more weight to information from certain people than others.

Table 1 shows the linkages between the groups, sub-groups and biases as categorized by Benson (2017). Note none of the biases discussed in this paper fit into the "What should we remember?" group.

Table 1. Groups, sub-groups and biases [4]

Group	Sub-group	Bias
We need to act fast	To avoid mistakes we aim to preserve autonomy and group status to avoid irreversible decisions	Status quo bias
	To get things done, we tend to complete things we've invested time & energy in	Sunk cost fallacy
Too much information	We notice when something has changed	Framing
		Anchoring
	We are drawn to details that confirm our own existing beliefs	Confirmation bias
Not enough meaning	We project our current mindset and assumptions onto the past and future	Hindsight bias
	We imagine things and people we're familiar with or fond of as better	Halo effect

2 How biases impact decision making

System 1 is the part of our mind that makes decisions quickly, and without active thought. It is doing this based on our past experiences and various cognitive biases that exist for each of us. This means that our biases have an effect on our decision making when our system 1 is in control.

2.1 Status quo bias

This is a tendency to want things to remain the same as they have been, an avoidance of change [5]. An everyday example here may be remaining with your current cellular phone provider even though a competitor may offer better rates or service. This type of bias can become an issue if the current way is not working or has unmanaged hazards in it.

2.2 Sunk cost fallacy

This is a tendency to want to continue with a course of action, even when evidence suggest otherwise, in an attempt to recoup losses [5]. It can occur when large sums of money or effort have been invested in something and it has not shown the expected rewards. There can be a tendency to invest more, even though evidence suggest that it will not be successful or indeed hazardous to continue. An everyday example here could be to 'double or nothing' in a game of chance. The odds of winning have not got better, but an attempt is made to win back previous losses by betting higher amounts. This type of bias can become an issue where it is potentially not safe to continue.

2.3 Framing

How information is presented can change how the information is interpreted [5]. An everyday example here may be that a yoghurt brand markets it product as 90% fat free, the corollary to this is that the product is 10% fat. If you are conscious about the amount of fat in your food you would more likely choose the 90% fat free rather than the 10% fat product. This type of bias can become an issue where data is presented in a misleading way. It may also have an opposite effect when some data is presented in a way to prime the mind so that it is more likely to make a favorable decision when caught up in the "thin slicing" stage of thinking [3].

2.4 Anchoring

This is based on placing more emphasis on the first information received, even if subsequent information shows it to be false [5]. An everyday example here might be when shopping and seeing a marked down price – 'was $200 now $150'. Seeing the first price of $200 may make you think that $150 is a good deal, because your expectation is anchored at $200. This type of bias can become an issue where the first piece of data is incorrect, and should be discounted from the decision making process.

2.5 Confirmation bias

This relates to placing more emphasis on data that supports your initial conclusion [5]. It can lead to the discounting of contrary data or searching for data that confirms your thoughts. An everyday example here is related to social media algorithms. We are presented with streams that typically confirm our own beliefs and we do not see counter information, which can lead us to believe that our thoughts are the same as the broader community. This type of bias can lead us to make the wrong decision because we do not accept data that disagrees with our initial thoughts.

2.6 Hindsight bias

This bias "is defined as the belief that an event is more predictable after it becomes known than it was before it became known" [6, p. 411]. Hindsight makes it difficult to understand why a person took the action they did when it led to an incident occurring. We are left asking questions like 'why did they do that; they must have known what would happen?' An everyday example here is preparing to go away for a camping trip, and on arrival realizing that an important piece of equipment was left behind, only to exclaim 'I knew I would forget something!' If this was actually known ahead of time it could have been prevented. This type of bias can lead us to believe that an incident could not happen to us, because would not make that kind of mistake that led to an incident.

2.7 Halo effect

This is a tendency to like everything about a particular person, even aspects of them you have not observed [2]. It can lead to following the advice of a person because you do not consider that they may be wrong. The corollary of this is the horns effect, where you do not believe the advice of a person because you do not believe they could be right. An everyday example of this may be thinking that your favorite actor must be a very nice person based on their attractive physical appearance, even though you have never met them. This type of bias can lead us to make an incorrect decision based on the opinion of others.

3 Biases in process safety incidents

There are several well know incidents where cognitive biases contributed to the event occurring. An incident has been selected for each bias, with the exception of hindsight bias, as the typical concern with hindsight bias is that it can inhibit learning from incidents after they have occurred and while this can result in a repeat incident, that has not been explored here.

The summary of incidents and their associate biases are contained in Table 2. This paper does not go into detail about the incidents, only mentioning the bias impacts. For more information on each incident refer to the references.

Table 2. Incident examples and their associated bias.

Bias	Incident	Example
Status quo bias	Columbia	Over the years that the space shuttle program had been running, NASA has normalized foam strikes as a part of launch. This because the status quo, and it was determined that they were not a safety of flight issue, and therefore there

		was no need to change the status quo [7].
Sunk cost fallacy	Bhopal	The Bhopal facility had a history of financial losses and equipment failures resulting in leaks. The facility had been shut down for a period of time and a decision was made to restart it to consume the feedstock, rather than dispose of it. During the final operation the toxic cloud was released. The decision to do the final produce run is an example of the sunk cost fallacy, continuing to run the facility, when it should have been shut down [8].
Framing	Challenger Space Shuttle	Temperature criteria had not been established for the solid rocket booster o-rings, but there was concern that the o-rings were brittle at low temperature. The initial framing suggested that the temperature effect was inconclusive, leading to concerns about the potential blow by of the o-rings being discarded [9].
Anchoring	Texas City Refinery	During the start up of the ISOM unit, the Raffinate Splitter Tower was filled with heated raffinate. The level indicator in the tower only covered the bottom section of the tower, and was also providing faulty readings. The operators became anchored to the level indicator results, not realizing the tower was in fact over filling [10].
Confirmation bias	Macondo	During the abandonment activities a cement barrier was installed. At the time it was believed that the installation went well and therefore the barrier would work. This led to test results being misinterpreted when assessing the barrier integrity. The belief that the barrier installation had been successful led to confirmation bias discounting the failed test results as incorrect [11].

Halo effect	Dreamworld	The Halo effect can exist when a junior or inexperienced worker is working at the direction of a more experienced person or being trained. As the junior person does not have the knowledge themselves, they will often do as they are told, deferring to the experience – the halo. The junior operator was told during her training she did not need to operate the emergency stop as that was the senior operator's role. When the raft became stuck on the rails no one hit the emergency stop, allowing the following raft to collide with it. Four people were killed when the second raft flipped as a result of the collision [12].

3.2 Hindsight bias

Hindsight bias often inhibits our ability to learn from an incident, because "the very outcome knowledge which gives us the feeling that we understand what the past was all about may prevent us from learning anything from it." [13, p. 311] In summary, we do not believe we would make the same mistakes. This phenomenon has led the IChemE Safety Centre to develop interactive case studies, where an attempt is made to remove the hindsight related to an incident, by telling the story without disclosing the outcome. Anecdotally these case studies appear to create an incident experience for the participant, without the consequences, which may assist with the learning being retained [14].

4 Ideas to avoid bias

The term often used for avoiding biases is called "debiasing". When considering the idea of system 1 and system 2 in our minds, to effectively debias we need to learn to engage system 2 to verify, correct and override where necessary our system 1 response. System 2 can be engaged by being asked questions that require thought, as this engages it to start to process. It has been found in studies that frowning while undertaking a task can increase the vigilance of system 2. This is because frowning usually occurs with cognitive strain, and this strain can be induced by frowning [2]. There are several ideas that can be tried to counter biases shown in Table 3.

Table 3. Ideas to counter bias

Bias	Ideas to counter it
Status quo bias	Appoint a person in the discussion to have the role of challenging the ideas, ensuring they have psychological safety to speak up without fear. This is akin to the "black hat" idea [15]. In this way there is someone free to ask questions about why the status quo exists as it is. The exploration of the status quo may highlight aspects that need changing.
Sunk cost fallacy	Engage someone from outside the issue, a coworker who is not tied to the outcome, to challenge what the next steps should be. They can help establish the pros and cons for each action, leading to a risk based decision on progression or not.
Framing	Reframe the data and discuss both aspects, identifying the counterfactual and assess both sides. From a priming aspect, it is possible to place subconscious thought in the mind prior to making decision. [3] An example might be that prior to undertaking a high risk task, the hazards and examples when things have previously gone wrong are specifically discussed. This act may prime the mind to look for and identify similar patterns.
Anchoring	When a decision needs to be reached, instead of a free discussion, have each person write down their initial ideas first. This way they have their thoughts that are not influenced by others during the discussion.
Confirmation bias	Engage someone from outside the issue, a coworker who is not tied to the outcome, to challenge what the data is telling you – do they reach the same conclusion without the background confirmation?
Halo effect	Attempting to slow down the thought process can assist with the time to look for more data to draw a conclusion from. If you are aware of the halo effect, you can try to think of different impressions of the person. This may be done by asking questions to try and establish a better understanding of them.
Hindsight bias	Attempt to understand the context in which decisions are made, not just the decision. This is useful during incident investigation, understand why an action was taken rather then what the action was. When we understand the context of

	decisions, we have the potential to see how it occurred instead of dismissing and thinking we would not have done it.

5 Conclusion

All actions and decisions we make are subject to biases, and we are often not aware of them. While most of the time this does not have an impact on the safety at a facility, sometimes the biases can lead to decisions which cause incidents. For this reason, it is important to understand the biases that may be present, as well as how our minds work with system 1 and system 2 and develop strategies to check if our system 1 decisions are likely to result in an incident.

6 References

[1] M. Haselton, N. D and A. PW, "The evolution of cognitive bias," in *The Handbook of Evolutionary Psychology*, Hoboken NJ, Wiley, 2005, pp. 724-746.

[2] D. Kahneman, Thinking, Fast and Slow, New York: Farrar, Straus and Giroux, LLC, 2011.

[3] M. Gladwell, Blink, Boston: Little, Brown and Company, 2005.

[4] B. Benson, "Cognitive Bias Codex," September 2017. [Online]. Available: https://www.visualcapitalist.com/wp-content/uploads/2017/09/cognitive-bias-infographic.html . [Accessed 19 July 2020].

[5] W. Hersing, "Managing cognitive bias in safety decision making: Application of emotional intelligence competencies," *Journal of Space Safety Engineering,* vol. 4, pp. 124-128, 2017.

[6] N. Roese and K. Vosh, "Hindsight bias," *Perspectives on Psychological Science,* pp. 411-426, 2012.

[7] Columbia Accident Investigation Board, "Columbia Accident Investigation Board," Columbia Accident Investigation Board, Washington DC, 2003.

[8] K. Bloch, Rethinking Bhopal, Amsterdam: Elsevier, 2016.

[9] Roger's Commission Report, "Report of the Presidential Commission on the Space Shuttle Challenger Accident," Presidential Commission on the Space Shuttle Challenger Accident, Washington DC, 1986.

[10] CSB, "Investigation Report Refinery Explosion and Fire," CSB, WashingtonDC, 2007.

[11] CSB, "Investigation Report Executive Summary Drilling Rig Explosion and Fire at the Macondo Well," CSB, Washington DC, 2016.

[12] Coroners Court of Queensland, "Inquest into the deaths of Kate Goodchild, Luke Dorsett, Cindy Low and Roozbeh Araghi at Dreamworld, October 2016," Queensland Courts, Southport, 2020.

[13] B. Fischhoff, "Hindsight ≠ foresight: the effect of outcome knowledge on judgement under uncertainity," *Quality and Safety in Health Care,* vol. 12, pp. 304-312, 2003.

[14] T. Kerin, "Accounting for hindsight bias: improving learning through interactive case studies," *Loss Prevention Bulletin,* no. 264, pp. 17-20, 2018.

[15] E. de Bono, Six thinking hats, Boston: Little, Brown and Company, 1985.

Unusual Hydrogen Explosions Due to Unanticipated Metal-Water Reactions

Robert Zalosh
Firexplo LLC
Wellesley, MA 02481
bzalosh@firexplo.com

Prepared for Presentation at
American Institute of Chemical Engineers
2020 Spring Meeting and 17th Global Congress on Process Safety
Virtual
April 18 - 22, 2021

Unusual Hydrogen Explosions Due to Unanticipated Metal-Water Reactions

Robert Zalosh
Firexplo LLC
Wellesley, MA 02481
bzalosh@firexplo.com

Keywords: hydrogen explosions, electrochemical reactions, metal-water reactions.

Abstract

Inadvertent generation of hydrogen from ambient temperature metal-water chemical and electrochemical reactions have been responsible for a surprising number of hydrogen explosions. This paper describes three sets of incidents involving different metals in applications that are normally considered free of explosion hazards. The first set of incidents involves wet alkaline incinerator ash with an aluminum content of at least two percent. The second set of incidents involves copper conductor electrochemical reactions in subsea electrical cable. The third set involves electrochemical corrosion reactions in galvanized dry sprinkler pipe. In all three applications, test data are discussed to elucidate the specific reactions and potential preventive and mitigation measures.

1 Introduction

Inadvertent generation of hydrogen from reactive powdered metal-water reactions and molten metal-water reactions is a well-known hazard in metal production and processing facilities. Prevention and protection measures for this hazard are described, among other places, in the NFPA 484 Standard on Combustible Metals [1]. However, inadvertent hydrogen generation from ambient temperature metal-water chemical and electrochemical reactions can also occur in many unanticipated settings and scenarios, and such scenarios have been responsible for a surprising number of hydrogen explosions.

Each of the following three sections of this paper provides details of one incident and then summarizes several other incidents for a given metal and application. It then describes the specific reactions likely responsible for the inadvertent hydrogen generation, including pertinent test data. The first set of incidents involve hydrogen generated from reactions of water with Municipal Solid Waste incinerator ash containing aluminum. The second type of metal-water electrochemical reaction involves copper cable in a seawater atmosphere. The third set of incidents entails hydrogen generation from galvanized steel corrosion reactions in dry pipe sprinkler systems.

2 Water-Incinerator Ash Reaction Incidents

2.1 Nortrader Cargo Vessel Explosion

A prime example of this type of incident is the January 2017 explosion on the dry cargo ship *Nortrader* that was the subject of a comprehensive investigation and report by the British Marine Accident Investigation Branch (MAIB) [2]. The *Nortrader* cargo of unprocessed incinerator bottom ash was generated in a nearby Plymouth, England power plant burning Municipal Solid Waste (MSW). Prior to the January 2017 incident, there had been 34 uneventful shipments of MSW bottom ash on 26 vessels similar in size to the 88 m length, nominal 1900 metric ton capacity *Nortrader*.

On January 12, 2017, over a period of about eight hours, 2,333 metric tons of wet MSW bottom ash were loaded into the *Nortrader's* 3409 m^3 cargo hold during steady heavy rain. After loading, the vessel left the dock and anchored overnight in Plymouth Sound while it waited for the weather to improve. During this time, the cargo hold ventilation fans and inlets were closed because of the heavy rain.

A large propagating explosion occurred the next day while the *Nortrader* was still at anchor. The ignition source for the explosion was the actuation of a stop button on an electrical panel situated in a ship compartment immediately forward of the large cargo hold. The ship's chief engineer, who was standing in front of the electrical panel suffered first degree burns to his face and second degree burns to his body, both hands, and lower extremities. The explosion then propagated into the cargo hold and lifted and bent all nine of the cargo hold steel latch covers, which had been held in place by latch cleats. The displaced hatch covers over the cargo hold the are visible in Figure 1.

Figure 1. *Nortrader* damaged cargo hold latch covers after explosions.

After the incident, the MSW from the Nortrader cargo hold was unloaded for inspection and testing. Employees of a consulting company retained by one of the ship's insurers drilled holes into the unloaded pile of MSW bottom ash and tested for the presence of flammable gas. They detected 1.4% hydrogen and 0.2 % carbon monoxide in the pile void spaces.

The MAIB hired another consulting company to acquire samples of the unloaded bottom ash for two types of water reactivity tests. One test series was UN Test N.5, the test method for substances which in contact with water emit flammable gases. It is described in section 33.4.1.4 (Annex E) of the UN Manual of Tests and Criteria [3]. The second test series involved an ad-hoc test with a much higher water-to-ash ratio (6:1) than in UN Test N.5 (1.5:1). Sample preparation for both tests entailed grinding the sample particulate to less than 250 micrometers for UN Test N.5, and to less than 1 mm for the ad-hoc reactivity test.

Several of the ash subsample water reactivity tests resulted in measured hydrogen production rates between 0.41 l/kg-hr and an average of 0.77 l/kg-hr. These rates are slightly less than the UN Test N.5 threshold of greater than 1 l/kg-hr of flammable gas to classify a material as a Division 4.3, substance which emits flammable gas when in contact with water [3]. Nevertheless, these hydrogen production rates indicate a potential for hydrogen accumulation in cargo holds and other enclosures with negligible ventilation and large quantities of bottom ash. Perhaps the UN Manual test criteria should be re-evaluated to provide a separate, less restrictive, category for this type of material that generates lower rates of flammable gas when in contact with large quantities of water.

The unventilated *Nortrader* cargo hold containing water-soaked bottom ash allowed the slowly generated hydrogen to accumulate over a period of about 19 hours while it was anchored in Plymouth Sound. An unsealed cargo hold light fixture access cover provided a path from the cargo hold to the adjacent ship compartment, which was also unventilated during this period. According to the MAIB report, the electrical panel in this compartment had an unsecured door, which allowed hydrogen to enter the panel interior. When an electrical circuit within the panel was energized and an arc or spark developed across a pair of contacts, the hydrogen-air mixture within the panel was ignited. The resulting explosion propagated from the panel into the ship compartment and then, via open light fixture cover, into the large cargo hold. An analysis by a MAIB contractor concluded that the hydrogen concentration in the cargo hold at the time of the explosion was in the range 6% to 9%, corresponding to a hydrogen volume between 114 m^3 and 171 m^3. The two explosions felt and heard by the *Nortrader* crew were the explosions in the electrical panel compartment and in the cargo hold itself.

2.2 MSW incinerator facility explosions

There have been several reported hydrogen explosions at MSW incinerator facilities. One such incident in Japan in 1995 is well documented [4,5,6]. A large amount of aluminum in the waste stream (9.6% Al) produced a clinker, i.e., a hard chunk of solidified molten metal and bottom ash, that was blocking flow in an ash discharge chute leading from the bottom of the incinerator to an ash hopper. Employees attempted to remove the clinker by striking it with a chisel and intermittently pouring water on it. This caused an explosion with ejected hot gas and ash burning three of the workers, one of whom subsequently died from the burn injuries [4,6]. The explosion was attributed to hydrogen formed from the aluminum-water reaction, with the subsequent hydrogen-air mixture in the chute being ignited by the chisel impacting the still hot clinker and ash.

A similar explosion occurred in a Japanese MSW incinerator in 1983 [6]. This explosion occurred in a dust bunker when an iron bar was used to crush some blocks of ash. The worker suffered severe injuries when he was thrown from the explosion site. The ash moisture content was about 10%, and there were significant amounts of aluminum in the ash.

Larson [7] briefly described a 2003 explosion in the basement ash storage bunker at a Swedish MSW incinerator due to hydrogen generation. Fortunately, there were no injuries, but the explosion damaged the bunker walls. Other incinerator facility explosions in Sweden have occurred during conveyor transport of warm wet bottom ash [8].

2.3 Other ash-water reaction explosions

Arm and Lindeburg [8] mention the occurrence of hydrogen explosions due to ash-water reactions in some Swedish ash storage facilities such as rock caverns and former oil wells. They also mention some explosion incidents associated with wetted fly ash in landfills.

There have been some flash fire and explosion incidents involving hydrogen generated from the ash aggregate used to make lightweight/foamed concrete. One such incident in 2009 in the UK occurred in a former dry well that had recently been filled with foamed concrete [9]. A grinder being used to cut reinforcing rods above the concrete ignited the hydrogen-air mixture that had accumulated there. This and other UK incidents involving foamed concrete and bricks caused the UK Highways Agency to ban the use of foamed concrete made with incinerator bottom ash.

2.4 Ash-water reactions producing hydrogen.

Although a variety of metals can react with water to produce hydrogen, most of the incinerator ash generated hydrogen has been found to be due to aluminum-water reactions. The primary reaction responsible for hydrogen generation is believed to be [10]:

$$2Al + 6H_2O \rightarrow 2Al(OH)_3 + 3H_2 \quad (1)$$

This reaction is promoted in an alkaline environment, with the major alkali source in MSW incinerator ash being calcium hydroxide. According to Nithiya et [10], the calcium hydroxide is consumed and regenerated in the following reactions, which combine to produce reaction (1).

$$2Al + 6H_2O + 2Ca(OH)_2 \rightarrow 2CaAl(OH)_5 + 3H_2 \quad (2)$$

$$2CaAl(OH)_5 \rightarrow 2Ca(OH)_2 + 2Al(OH)_3 \quad (3)$$

The presence of these compounds in ash-water reaction products has been confirmed by Nithiya et al [10], and the promoting effect of pH values > 10 in ash has been demonstrated in test data reviewed by Larsen [7] and others. Several researchers have reported significantly greater hydrogen generation rates when the ash is mixed with alkaline solutions compared to mixing with distilled water. Nithiya et al [10] cite research showing that the alkaline ash conditions cause the oxide layer on the aluminum surface to be perforated and dissolved, allowing direct contact and reaction of aluminum and water.

As implied by reactions (1) and (2), hydrogen generation rates from ash-water reactions increase with increasing aluminum content and increasing pH of the ash. The reactions have also been shown to be faster for freshly generated warm ash, and to decay after several days of ash exposure to ambient air (allowing greater oxide layer formation). These effects produce a wide variation in reported hydrogen generation yields obtained in tests with MSW ash samples.

Besides hydrogen generation rate data, reported test data include total hydrogen generated per unit mass of ash. The hydrogen yield data show the yield increasing with increasing ash aluminum content as well as other factors mentioned above. Yield data have been compiled from a variety of sources including data for bottom ash from Japanese incinerators [10], fly ash from Japanese incinerators [6], and both bottom ash and fly ash from Swedish incinerators [8]. The hydrogen generation data cumulative distribution for ambient temperature reactions obtained by mixing with distilled water are shown in Figure 2. The median hydrogen yield is 2.4 liter-H_2 per kg-ash for bottom ash and 10 l/kg-ash for fly ash. The highest measured hydrogen yields are 24 l/kg-ash for bottom ash and 37 l/kg-ash for fly ash.

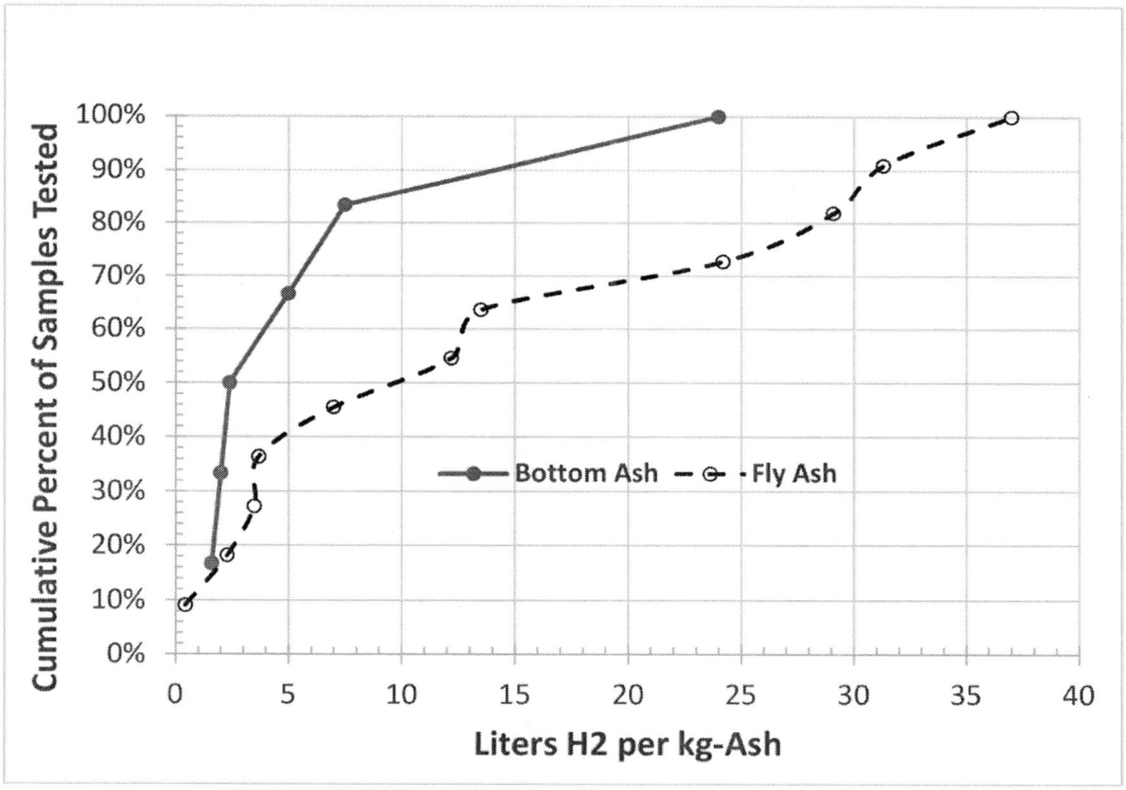

Figure 2. Hydrogen generation data cumulative distribution for bottom ash and fly ash reactions with water.

The data shown in Figure 2 can potentially be used to provide a risk-based estimate of the hydrogen accumulation hazard potential in MSW incinerator ash bunkers and storage and transportation facilities. Daily generation data reported in the cited references show that most of the hydrogen is generated in the first few days after ash formation and contact with copious quantities of water. (Based on the test methods cited above, copious water implies having a water/solid ratio of about two or greater at some point during handling and storage.) Therefore, this data can also be used as a basis for estimating MSW ash storage facility ventilation requirements to prevent flammable mixture formation. These requirements are not needed for ash obtained from other fuels providing those fuels do not have aluminum contents of about 2 percent or higher.

3 Electric Cable – Seawater Reaction Hydrogen Incidents

3.1 Offshore Junction Box Explosions

Offshore oil and gas exploration and production platforms have many electrical junction boxes exposed to sea air and aerosols that produce a saline environment within the box. Several of these boxes have experienced explosions that damage the box. Incident investigations have determined that hydrogen generation within the box produced a flammable atmosphere that was probably ignited by an arc, spark, or electrostatic discharge.

One junction box explosion incident [11] occurred in a box installed above the boat landing level of an unmanned platform in a tropical environment. The box had an IEC 60079 [12] designation "Ex e", meaning that it is intended for normally non-sparking electrical equipment with a rated voltage no greater than 11 kV, to be installed in a potentially flammable gas atmosphere. This particular box contained a 6.6 kV submarine power cable, and had breather ventilation.

During a maintenance crew visit about a year after the cable was first energized, the junction box hinged door was observed to be bulged outward and blown open, as can be seen in Figure 3. There was no interior damage or char deposits and the electrical connections remained intact. The fact that the box did not withstand the explosion is not surprising since it was not constructed or rated as an "Ex d" box.

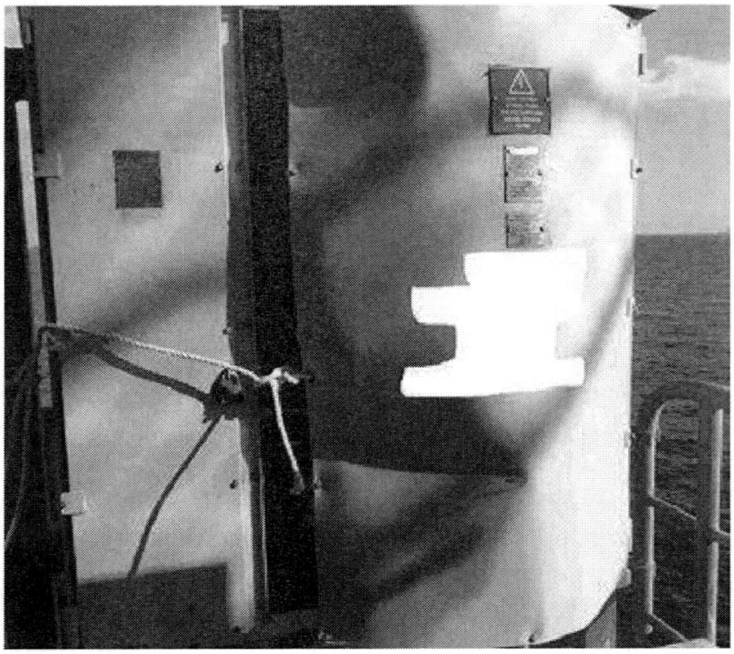

Figure 3. Damaged junction box door on offshore platform [11].

There have been at least two similar reported junction box explosion incidents at other platforms and locations. After one of those incidents, flammable gas was detected in the box [11].

Hydrogen generation from the copper conductor within seawater cables is a recognized phenomenon in subsea electrical cables with the generation and associated copper loss being more pronounced when there is cable insulation damage. This is illustrated in Figure 4, which shows the hydrogen bubbles originating from damaged cables in simulated seawater [13]. Hydrogen is also formed from seawater reactions with the copper and steel components of optical fiber cable [14]. Some seawater cable manufacturers provide hydrogen absorbing materials in their cables to prevent the hydrogen from being released from the cables.

Figure 4. Hydrogen bubbles from a damaged cable in simulated seawater [13]

3.2 Buried Cable Seawater Reaction Hydrogen Incident

An unusual electric cable-seawater reaction hydrogen explosion occurred at a Rhode Island beach in July 2015 [15]. The hydrogen was generated by the seawater corrosion of an abandoned electrical cable that had been used until 2007 to power Coast Guard navigational lights on rock jetties near the beach. People at the beach were not aware of the cable which was buried in the sand. When a hydrogen-air mixture was somehow ignited, the explosion blew a woman out of her beach chair. She suffered broken ribs and bruises upon landing on a jetty about 10 feet away.

Figure 5 shows a portion of the cable after the sand that had obscured it was removed. University of Rhode Island geochemists tested sand samples from the area and measured hydrogen concentrations as high as 5%, i.e., slightly above the Lower Flammable Limit. They also observed that the corroded cable did not have a green patina, indicating that the corrosion was due to oxide, rather than chloride formation.

Figure 5. Corroded electric cable under beach sand [15]

3.3 Seawater Induced Copper Hydrogen Evolution Reactions

Copper exposure to a seawater alkaline environment promotes hydrogen evolution via two different electrochemical reactions. First, copper serves as a catalyst for the cathodic reaction

$$2H_2O + 2e^- \rightarrow H_2 + 2[OH]^- \quad (4)$$

Sherif and Park [17] maintain that this reaction is more likely to occur in de-aerated solutions, whereas aerated NaCl solutions are more likely to produce the water and oxygen reduction reaction

$$O_2 + 2H_2O + 4e^- \rightarrow 4[OH]^- \quad (5)$$

However, Martins et al, [18] have shown that reaction (4) can produce copious amounts of hydrogen when the polycrystalline copper surface allows the formation of a copper-hydroxide activated complex. They demonstrate that copper surface treatment via polishing and annealing allows this activated complex to form and render the copper surface more hydrophilic.

The second seawater induced hydrogen evolution pathway is the following direct copper reaction with water [17]

$$2Cu + H_2O \rightarrow Cu_2O + 2H^+ + 2e^- \rightarrow Cu_2O + H_2 \quad (6)$$

Reaction (6) is suppressed in the absence of seawater because of the compact film of the cuprous oxide reaction product. Upon seawater exposure, chloride ions cause the formation of a cuprous chloride ion that partially dissolves the cuprous oxide film and forms the more porous Cu_2O and layer of atacamite ($Cu_2(OH)_3$- Cl), which readily admits more chlorine ions to reach the copper underneath it [17, 19]. The dissolution/porosity of the oxide film allows the oxidation reaction to continue and also results in further copper dissolution. Thus, copper corrosion can be often seen in the form of the green patina due to the cuprous chloride, and sometimes seen merely as a loss of copper without any discoloration. The latter was reported to be the case in the buried cable beach incident summarized above [16].

Reported test data on copper seawater corrosion rates can be used in conjunction with Reaction (6) to quantify the corresponding expected hydrogen generation rates. Corrosion rate data used for this purpose was obtained from copper exposure laboratory tests with natural and artificial and aerated seawater [19, 20], and from field tests with copper samples in seawater and sea aerosols in North Atlantic and tropical environments [21, 22]. The calculated hydrogen generation rates for exposures of several months varied from about 2 cc-H_2-per-hr-m^2-Cu to almost 14 cc-H_2 per hr-m^2. The cumulative distribution of the calculated results is shown in Figure 6. These results can be used to determine ventilation rates needed to prevent flammable hydrogen-air mixtures in electrical junction boxes or other small electrical enclosures in a seawater or sea atmosphere environment.

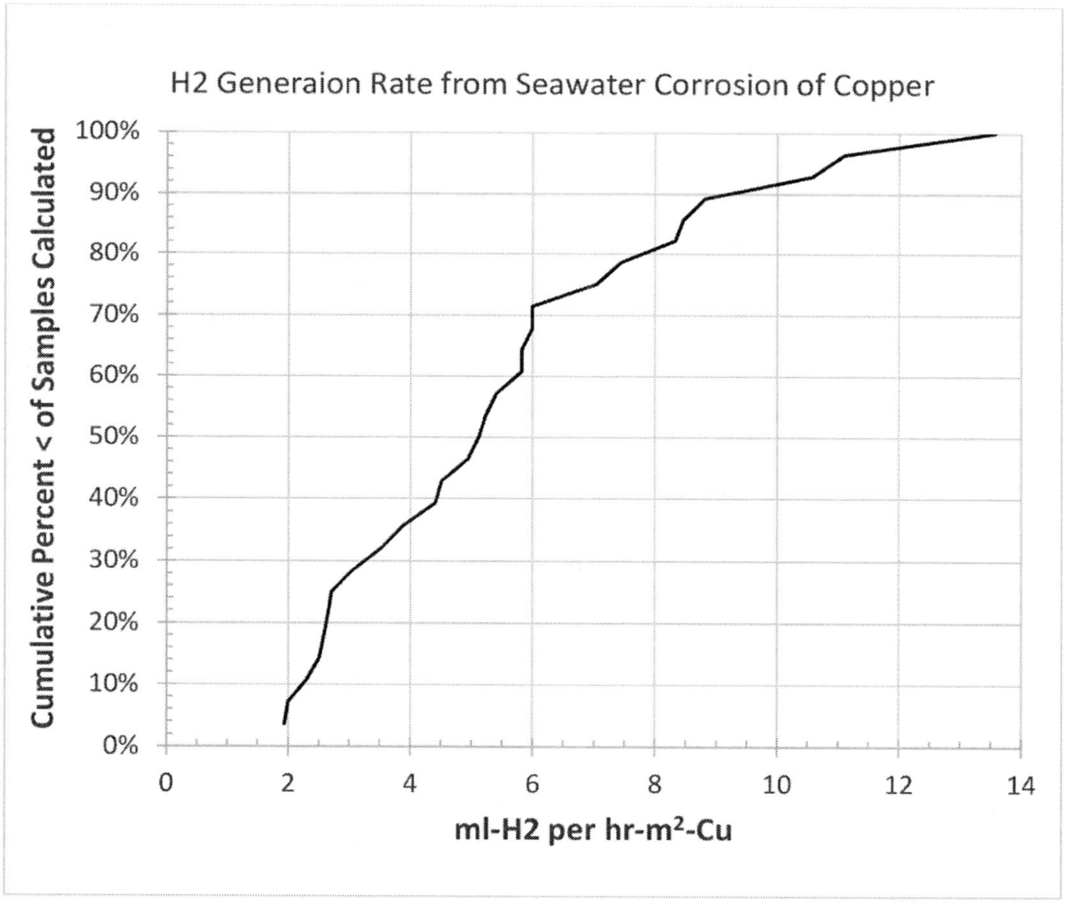

Figure 6. Calculated hydrogen generation rates from seawater exposure of copper.

4 Galvanized Sprinkler Piping Hydrogen Incidents

4.1 Helsinki Theater Sprinkler Room Explosion

On June 29, 2020, a sprinkler contractor was conducting routine maintenance on a dry pipe sprinkler system in the basement of a Helsinki theater. Shortly after the contractor opened the pipe, an explosion occurred that caused burn injuries to the contractor, who required hospitalization. The explosion caused minor damage to the stairwell adjacent to the sprinkler room [23].

An investigation by the Finnish Safety and Chemical Agency (Tukes) determined that the hydrogen was generated from a corrosive reaction involving zinc, water, and air within the galvanized steel pipe [24]. It was the first incident of its kind in Finland, but Tukes was sufficiently concerned about possible future incidents to issue a ban on the use of zinc-coated pipes in new sprinkler systems. Tukes also issued new requirements for maintenance involving opening of sprinkler piping; ventilation and a hydrogen detector are now required.

4.2 Other Sprinkler Pipe Hydrogen Fire and Explosion Incidents

Although the Helsinki theater explosion was the first incident in Finland involving hydrogen generated in galvanized sprinkler piping, there have been several such incidents in other Scandinavian countries, where galvanized dry pipe sprinkler systems are common. The following is a list of some European incidents.

- A 2014 incident occurred at a Danish industrial company when a sprinkler contractor was burned by a hydrogen flame vented from sprinkler piping while it was being drained [22].
- A second 2014 incident in Denmark; this one involved a hydrogen release from a pressurized water tank that had recently been drained. A sprinkler contractor suffered burn injuries from the resulting room explosion which may have been ignited by sparks from the contractors' tools [25].
- There was an explosion in a sprinkler fire pumphouse in France in October 2016 after a sprinkler contractor had drained the sprinkler piping in preparation for doing work on the system. The explosion occurred about 15 minutes after the drain was opened and the contractor left the pumphouse. Fortunately, there were no injuries, but the pumphouse was destroyed [26].

There have also been similar reported incidents in Japan. Yamate et al. [24] mention ignitions occurring when wet galvanized pipe have been cut or subjected to grinding during sprinkler system renovations.

4.3 Hydrogen generation reactions in sprinkler piping

The electrochemical reactions responsible for hydrogen generation in galvanized sprinkler pipe are analogous to the reactions described above for copper conductor exposure to seawater. The reactions described by Yamate et al. [27] are:

$$2H_2O \rightarrow 2H^+ + 2[OH]^- \quad (7)$$

$$2H^{+} + 2e^{-} \rightarrow 2H \rightarrow H_2 \quad (8)$$

$$Zn \rightarrow Zn^{2+} + 2e^{-} \quad (9)$$

$$Zn^{2+} + 2[OH]^{-} \rightarrow Zn(OH)_2 \quad (10)$$

Reactions (9) and (10) produce the zinc corrosion that results in formation of the white zinc hydroxide layer seen below the indicated water line in Figure 7. Su and Fuller [28] did not observe this corrosion when the air above the water was replaced by compressed nitrogen. This suggests that oxygen is required for accelerated corrosion in galvanized pipe, and that oxygen-water electrochemical reaction (5) is a key aspect of both corrosion and hydrogen production in galvanized dry pipe sprinkler systems in which there is residual water together with air in the piping. This often occurs when the piping is not completely drained following a dry pipe valve trip test or inadvertent actuation.

Figure 7 Zinc corrosion produced in 6 weeks in half-filled pipe [25]

Yamate et al. [24] have shown that rapid hydrogen generation can also occur from galvanic corrosion associated with water filled galvanized sprinkler pipe in contact with another metal component such as a stainless steel flexible pipe section. Gas samples from the galvanic coupling pipe revealed a hydrogen concentration of 20% and significant concentrations of zinc in the water. They also obtained visual evidence of hydrogen bubbles leading to a flammable atmosphere when the piping was opened, as shown in Figure 8.

It foams. It ignited.

Figure 8 Hydrogen bubbles and flames in opened galvanic coupled piping [24]

Although the hydrogen generation incidents and accelerated corrosion observations have led to some bans on galvanized sprinkler piping, FM Global [28] notes that corrosion products in ordinary black carbon steel dry piping have been responsible for the vast majority of their fire losses due to obstructed sprinkler pipes. Therefore, FM Global continues to recommend use of galvanized steel pipe for dry and pre-action sprinkler systems. In view of the exposure test results noted above, FM Global recommends use of compressed nitrogen, rather than just air, in these dry pipe systems. This is different from the recommendations and requirements of many other organizations, as summarized in a recent Sprinkler Age article [29].

5 Conclusions

Metal-water chemical and electrochemical reactions have been responsible for hydrogen generation and subsequent ignition of hydrogen-air mixtures in many explosion incidents. Many incinerator ash-water incidents have occurred when the ash had a substantial aluminum and alkaline content. A compilation of test data has led to a statistical distribution of the expected hydrogen generation per unit mass of ash, with the median rate being greater for fly ash than for bottom ash.

Copper-seawater electrochemical reactions in offshore facility electrical junction boxes have led to hydrogen accumulations and subsequent ignitions associated with electrical connector arcs and sparks. Copper corrosion test data together with reported electrochemical reactions have allowed calculations of hydrogen generation rates per unit of copper cable surface area. Special cable features, such as hydrogen absorption additives, and flameproof junction box construction can help prevent and mitigate the consequences of these incidents.

Electrochemical corrosion reactions in galvanized dry pipe sprinkler systems containing residual water have led to many reported hydrogen fire and explosion incidents, particularly in Scandinavia, where there are many of these sprinkler systems. Testing has shown that the accelerated corrosion in galvanized dry pipe can be greatly reduced by replacing air in the pipe with compressed nitrogen, and by avoiding the galvanic corrosion resulting from contact between dissimilar metal components in sprinkler piping.

6 References.

[1] NFPA 484 Standard on Combustible Metals, National Fire Protection Association, 2019.

[2] Marine Accident Investigation Branch, Serious Marine Casualty Report Number 26/2017, December 2017.

[3] UN, Recommendations on the transport of dangerous goods, Manual of tests and criteria, United Nations, 5th Edition, 2009.

[4] https://h2tools.org/lessons/hydrogen-gas-explosion-municipal-refuse-incineration-facility. Accessed on 2/8/2021.

[5] http://www.shippai.org/fkd/en/cfen/CC1200025.html. Accessed 2/8/2021.

[6] S. Mizutani,, S. Sakai,, and H. Takatsuki, Investigation of hydrogen generation from municipal solid waste incineration fly ash, J Mater Cycles Waste Manag (2000) 2:16–23.

[7] R. Larsen, Energy recovery of metallic aluminium in MSWI bottom ash, Umea University M.S. Thesis, 2014.

[8] M. Arm, and J. Lindeberg, Gas generation in incinerator ash, 2015, Available at https://www.researchgate.net/publication/266467653. Accessed 2/8/2021.

[9] UK Highways Agency Interim Advice Notice 127/09, Appendix B: The Use of Foamed Concrete, October 2009.

[10] A. Nithiya, A.Saffarzadeh, and T. Shimaoka, Hydrogen gas generation from metal aluminum-water interaction in municipal solid waste incineration (MSWI) bottom ash, Waste Management 73 (2018) 342–350.

[11] Rupturing of 6.6kV Power Junction Box installed at an offshore oil production Well Head Tower, https://www.eng-tips.com/viewthread.dm?qid=434654. Accessed 2/12/2021.

[12] IEC 60079-7, Explosive atmospheres - Part 7: Equipment protection by increased safety "e", IEC, 2015.

[13] N. Douglas, Copper Loss from Electrical Conductors in a Subsea Environment as a Result of Low Insulation Resistance, Oilfield Technology, January 2019. Available at https://www.researchgate.net/publication/331951501.

[14] ITU-T Recommendation L.27, Method for estimating the concentration of hydrogen in optical fibre cables, International Telecommunication Union, 1997.

[15] State: Hydrogen from corrosion of copper cable caused Salty Brine beach blast, Providence Journal, July 24, 2015.

[16] Case Closed: Hydrogen Gas Combustion Caused Salty Brine Beach Blast, National Public Radio, July 24, 2015. Accessed 2/12/2021.

[17] E-S. Sherif, and S-M. Park, Inhibition of Copper Corrosion in 3.0% NaCl Solution by N-Phenyl-1,4-phenylenediamine, Journal of The Electrochemical Society, 152 10 B428-B433 (2005).

[18] P. Martins, et al, Hydrogen evolution reaction on copper: Promoting water dissociation by tuning the surface oxophilicity, Electrochemistry Communications 100 (2019) 30–33.

[19] E-S Sherif, R. Erasmus, and J. Comins, Corrosion of copper in aerated synthetic sea water solutions and its inhibition by 3-amino-1,2,4-triazole, Journal of Colloid and Interphase Science, 309 (2007) 470–477.

[20] F. Mansfield, et al., The Corrosion Behavior of Copper Alloys, Stainless Steels, and Titanium in Seawater, Corrosion Science, Vol. 36, No. 12, pp. 2063-2095, 1994.

[21] A. Drach, et al., Field studies of corrosion behaviour of copper alloys in natural seawater, Corrosion Science 76 (2013) 453–464.

[22] L. Nunez, et al, Corrosion of copper in seawater and its aerosols in a tropical island, Corrosion Science 47 (2005) 461–484.

[23] Explosion in Helsinki City Theater, https://www.foreigner.fi/content/print/explosion-in-helsinki-theater-at-least-one-person-in-hospital/20200629150657669l. Accessed 11/17/2020.

[24] Hydrogen explosion caused the Helsinki City Theatre incident, https://tukes.fi/en/-/hydrogen-explosion-caused-the-helsinki-city-theatre-incident. Accessed 2/14/2021.

[25] Explosions in sprinkler system probably caused by hydrogen gas, https://dpa-e.eu/explosions-in-sprinkler-system-probably-caused-by-hydrogen-gas/ Accessed 2/14/2021.

[26] Internally Galvanized Steel Pipe Blamed for Explosions, https://www.complianceonegroup.com/ lifesafety/ internally-galvanized-steel-pipe-blamed-for-explosions. Accessed 2/14/2021.

[27] T. Yamate, et al., Experimental Study on Hydrogen Generation Phenomenon in Fire Piping System in Buildings, Corrosion Engineering, 59, 436-439, 2010.

[28] P. Su, and D. Fuller, Corrosion and Corrosion Mitigation in Fire Protection Systems, FM Global Report, July 2014.

[29] Galvanized Steel Sprinkler Piping, Sprinkler Age, November 2018, Available at https:// sprinkle rage.com/ galvanized-steel-sprinkler-piping/ Accessed 2/15/2021.

619874 Port of Beirut Lebanon Explosion – Analysis and Impacts

Abstract:

On 4 August 2020, 2,750 tons of ammonium nitrate stored at the Port of Beirut (POB), Lebanon, exploded, reportedly resulting in over 190 deaths, several thousand injuries, $10-15 billion USD in property damage, and leaving an estimated 300,000 people homeless. The cargo of ammonium nitrate had been stored in Warehouse 12 at the POB, after having been confiscated by the Lebanese authorities from an abandoned ship, the MV *Rhosus,* in September, 2013. It is assumed that the ammonium nitrate was stored in the warehouse without proper security and safety measures, and without consideration for the need for segregation from other incompatible materials. From recorded videos of the blast, the explosion was preceded by a fire in the warehouse, but the exact cause of the detonation is still under investigation. This was one of the largest non-nuclear explosions in modern history and the lessons learned have widespread implications for facility siting, the handling of explosives and other hazardous materials, the management of chemicals in transportation, emergency management and recovery, and building resilience for large-scale emergencies.

In response to this event, AcuTech participated as part of an assessment team, and was tasked with examining current and previous practices for the handling of chemicals and hazardous materials at the POB, taking an all-hazards approach to examine safety, security, and emergency management concerns in a unified manner.

This paper examines the multifaceted impacts of the blast, the most critical factors that may have contributed to the explosion, and best practices and lessons learned that are applicable not only to ports and facilities handling ammonium nitrate, but all facilities handling hazardous materials operating in challenging urban environments, and which maintain complex interdependencies with critical infrastructure.

AUTHOR INDEX